Turbulent Jets

A Point-Source Method for Hydraulic Engineering

Bidya Sagar Pani

FOUNDATION® BOOKS

Delhi • Bengaluru • Mumbai • Kolkata • Chennai • Hyderabad
Pune • Thiruvananthapuram

Published by
Cambridge University Press India Pvt. Ltd.
under the imprint of Foundation Books
Cambridge House, 4381/4 Ansari Road, Daryaganj, **New Delhi** 110 002

Cambridge University Press India Pvt. Ltd.
C-22, C-Block, Brigade M.M., K.R. Road, Jayanagar, **Bengaluru** 560 070
Plot No. 80, Service Industries, Shirvane, Sector-1, Nerul, **Navi Mumbai** 400 706
10 Raja Subodh Mullick Square, 2nd Floor, **Kolkata** 700 013
21/1 (New No. 49), 1st Floor, Model School Road, Thousand Lights, **Chennai** 600 006
House No. 3-5-874/6/4, (Near Apollo Hospital), Hyderguda, **Hyderabad** 500 029
Agarwal Pride, 'A' Wing, 1308 Kasba Peth, Near Surya Hospital, **Pune** 411 011
T.C. 25/ 2731 Lukes Lane, Ambujavilasam Road, **Thiruvananthapuram** 695 001

First Published 2012

ISBN 978-81-7596-928-5

Published by Manas Saikia for Cambridge University Press India Pvt. Ltd. and printed at Raj Press, R-3 Inderpuri, New Delhi 110012

Contents

List of Figures

List of Tables

Preface

Turbulent jets and plumes occur in very many engineering applications. It is exceedingly important to predict how effluents mix and disperse in the environment. Many books on jets and plumes have been written in the past, each dealing with the subject matter from a different perspective. To cite an example, the recent book by Lee and Chu (2002) follows a Lagrangian approach. The current book is an outcome of over three decades of teaching and research by the author in the field of turbulent jets. It contrasts and complements the subject matter of other books.

The theme of the book is based on a simple momentum transport model proposed by Reichardt in the early forties of the twentieth century. The original model of Reichardt has been modified, and the point-source technique has been comprehensively developed in this book. The author has explained how the point-source concept and the method of superposition can be applied to many environmental and hydraulic engineering problems.

The new theoretical framework has been supported by many experimental observations. Some of the observations are old, and the others are of recent origin based on modern technologies such as LDA, PIV, LIF and video-imaging. For the sake of continuity, results of experimental investigations, carried out during the formative years of research on jets, have been included in the first few chapters.

The salient features of this book are the following:

- Theoretical expressions for the distribution of velocity, temperature and tracer concentration of a single three-dimensional outlet and the conventional round jet have been developed based on the point-source method. They provide a sound basis to analyse flow from multiple outlets.
- A new hypothesis, supported by experimental results, has been proposed to describe the excess momentum flux of a coflowing jet by a Gaussian distribution.

- A link factor has been introduced to generalise the application of Reichardt's hypothesis for a jet diffusing in a coflowing ambient.
- A theoretical framework has been developed to predict the velocity and tracer concentration field downstream of multiport diffusers, discharging effluents into rivers and streams.
- The effects of the free surface and the channel bottom have been incorporated in the model, which was not feasible earlier sans the superposition technique.
- The decay of the maximum velocity of single and multiple wall jets has been predicted based on the method of superposition by treating the wall as a frictionless reflector.
- The point-source technique has been applied to estimate the energy loss in a sewer junction box and in a jet flocculator of a water treatment plant.
- A single spread coefficient has been adopted for the near-field and the far-field flow. This novel idea overcomes many of the limitations of existing integral techniques.
- The new approach offers a more realistic estimate of the length of the potential core and a proper interpretation of experimental observations in the developing zone of a jet.
- The need to account for virtual origin correction has been eliminated.

For a beginner, the task of synthesising the vast available literature on jets, involving varied notations and normalisation scales, is difficult. This book has made an attempt to unify the underlying principles. Worked out examples have been provided to demonstrate the method of superposition. Derivations have been included where necessary. For the purpose of illustration, different types of jet flows are simulated making use of Flow3D CFD software, and the mean velocity results are compared with the ones predicted from the point-source method.

Chapter 1 gives a brief introduction on the application of jets to different problems in environmental and hydraulic engineering. Advantages of simple turbulence models for analysis of two-dimensional jet boundary layers have been elaborated in Chapter 2. Chapter 3 deals with round free jets, whose spreading properties can be applied to three-dimensional jets. Subsequently, the main theme of this book, Reichardt's hypothesis and the analysis built on the point-source concept are introduced in Chapter 4.

The application of the point-source to three-dimensional free jets in a stagnant ambient is discussed in detail in Chapter 5. Self-similarity of the excess-momentum flux and the method of superposition are the cornerstone of the theoretical framework adopted in this book. These are used to predict the velocity, concentration of tracer and dilution downstream of a jet in a coflowing ambient current and are presented in Chapter 6. The case of multiple jets, both in a stagnant and a coflowing stream, is dealt with in Chapter 7. The flow from a single outlet forms the basis for superposition. The effects of the channel bottom and the free surface are also accounted for in the model.

Two important issues, the length of the potential core and the location of the virtual origin, have been addressed in Chapter 8. The length of the potential core based on the point-source approach seems to be more realistic.

The free jet theory has been used in Chapter 9 to estimate the energy loss in a sewer junction-box. The efficacy of a jet flocculator for removal of turbidity from raw water supplied to a treatment plant is discussed in Chapter 10. Due consideration is given to the siting of the outflow section for achieving optimal performance of the flocculation chamber.

Wall jets have many applications in hydraulic engineering and are a distinct class by themselves. The point-source technique and the method of superposition have been extended to the case of wall jets. The wall is treated as a frictionless reflector and, through superposition, the decay of the maximum velocity and the concentration of different types of jets have been predicted. Chapter 11 gives a brief account of the different types of wall jets, inclusive of the effects of coflowing and crossflowing ambient streams.

I sincerely acknowledge the willingness of Dr Paul Chu and Dr H.J. Wang for agreeing to use their experimental data/plots on coflowing jets. The suggestions offered by Professor Joseph Lee during the course of writing this book, especially on coflowing jets, were very useful. I wish to thank him for lending support for my visits to the University of Hong Kong, which helped me to complete the manuscript. The support received from Professor N. Rajaratnam, my former PhD supervisor, is acknowledged.

My wife Sarama always provided me with the requisite encouragement to complete the manuscript. She took pains to go through the entire write-up and offered suggestions concerning the style of presentation. Without her input, the book might not have come to the present form, which I believe will yield the desired results. I shall consider my efforts worthwhile if the readers, especially the beginners in this subject, learn and benefit from the content of this book.

The figures appearing in this book have been prepared in the electronic form by two of my former students at IIT Bombay and two graduate students working in the Croucher Laboratory of Environmental Hydraulics, The University of Hong Kong. Their help is very much appreciated.

I wish to thank the following publishers for according copyright permission to reproduce and reuse figures from their publications:

IIT Bombay (PhD thesis)
MIT Press (*The Theory of Turbulent Jets*, Abramovich, 1963)

American Society of Mechanical Engineers (Trans. ASME, *Journal of Fluids Engineering*)
American Society of Civil Engineering (Proc. and Trans. ASCE)
American Institute of Aeronautics and Astronautics (*Journal of AIAA*)
Cambridge University Press (*Journal of Fluid Mechanics*)
Springer Verlag (*Applied Scientific Research, Experiments in Fluids*)
Institution of Civil Engineers, UK, London
Elsevier Scientific Publishing Company (*Turbulent Jets*, Rajaratnam, 1976)
International Association for Hydro-Environment Engineering and Research (IAHR, *Journal of Hydraulic Research*).

Bidya Sagar Pani

1 Jets in Hydraulic Engineering

1.1 Features of Jets

A jet by definition is a momentum-dominated flow from a continuous source. The Reynolds number of flow, $R_e (= \frac{U_0 d}{\nu})$, determines the nature of the flow, that is, laminar or turbulent. In most practical problems if the Reynolds number of flow, in terms of the outlet diameter, d, and efflux velocity, U_0, exceeds a value of 2000, then the jet is considered turbulent (Daily and Harleman, 1973).

Shear flows generate turbulence, and turbulence causes shear in the fluid flow. Therefore, turbulent motion in a shear flow has the property of attaining self-sustenance. Free turbulence occurs when shear flows persist at the interface between fluid zones. A typical example of the free turbulence is the spreading of a round jet from a nozzle into an ambient fluid of the same kind. In contrast, if the shear flow occurs very close to the solid boundary, the turbulence generated therefrom is classified as wall turbulence. In wall jets, the turbulence is a consequence of the wall shear. Generally, the analysis of wall jets is somewhat more difficult as the viscous stresses in the laminar sublayer adjacent to the solid boundary have to be retained in the governing equation of motion.

A boundary layer type of approximation has been adopted to simplify the governing equations of motion of free and wall jets. In addition, the velocity distribution u versus y (or u versus r) also exhibits 'self-similarity'. In the past, the simplified governing equations of motion for free shear flows were solved for the mean velocity distribution based on the principle of similarity. As an alternative, integral techniques are very often used because of their simplicity. In the integral method, a suitable velocity distribution is assumed, rather than determining the shape of the distribution. The integral techniques are versatile and most of the analysis discussed in this book is based on this approach.

All types of jets included in this book are considered as turbulent, and the diffusion process is independent of the Reynolds number. As there is no difference in density between the jet and the ambient fluid, gravity plays no role in the analysis. Also, the outcome of experiments on jets will be the same, irrespective of the fluid medium, that is, air or water.

In the available literature on the diffusion of jets, the case of a round jet emanating from a circular outlet is often known as the axially symmetric jet. There are instances when the round jet has also been inappropriately called a three-dimensional jet. In the present context, a free jet issuing from outlet shapes such as a square, a rectangle, an ellipse or even a triangle will be designated a three-dimensional jet. Interestingly, flow from a circular outlet too becomes three dimensional in nature in the presence of a wall adjacent to it. Thus, all wall jets, other than the plane slot case, are classified as three-dimensional wall jets.

1.2 Applications of Jets

Turbulent jets are used extensively in various disciplines such as aerospace engineering, chemical industry, mechanical engineering and civil and environmental engineering. One of the primary purposes of using a jet is to achieve mixing of chemicals and effluents in an efficient fashion. Often, jets are also used for dissipating excess energy of the flow, flocculation in water treatment plants, cleaning of gravel beds of rivers for fish spawning, cooling of surfaces, heating and ventilation of buildings, thrust augmentation and cutting of material. The examples quoted in this book, however, focus on the application of jets in civil and environmental engineering, though similar principles are equally applicable to other branches of engineering. Typical examples of turbulent jets as used in hydraulic and environmental engineering are briefly discussed below.

1.3 Some Examples of Jets

Schematic layouts of some types of jets used in practice are shown in Fig 1.1.

The flow at the toe of the bucket, shown in case (a), is more complex than that shown in the other examples due to the presence of a roller. The roller mechanism pushes the bed material towards the toe, and excessive scour is inhibited. A detailed knowledge of the flow pattern, especially the one appearing in and around the roller, can help arrive at a suitable geometry of the bucket.

Case (b) shows how effluents are often discharged into streams or ocean currents through a single or multiport diffuser. The goal is to mix the effluents

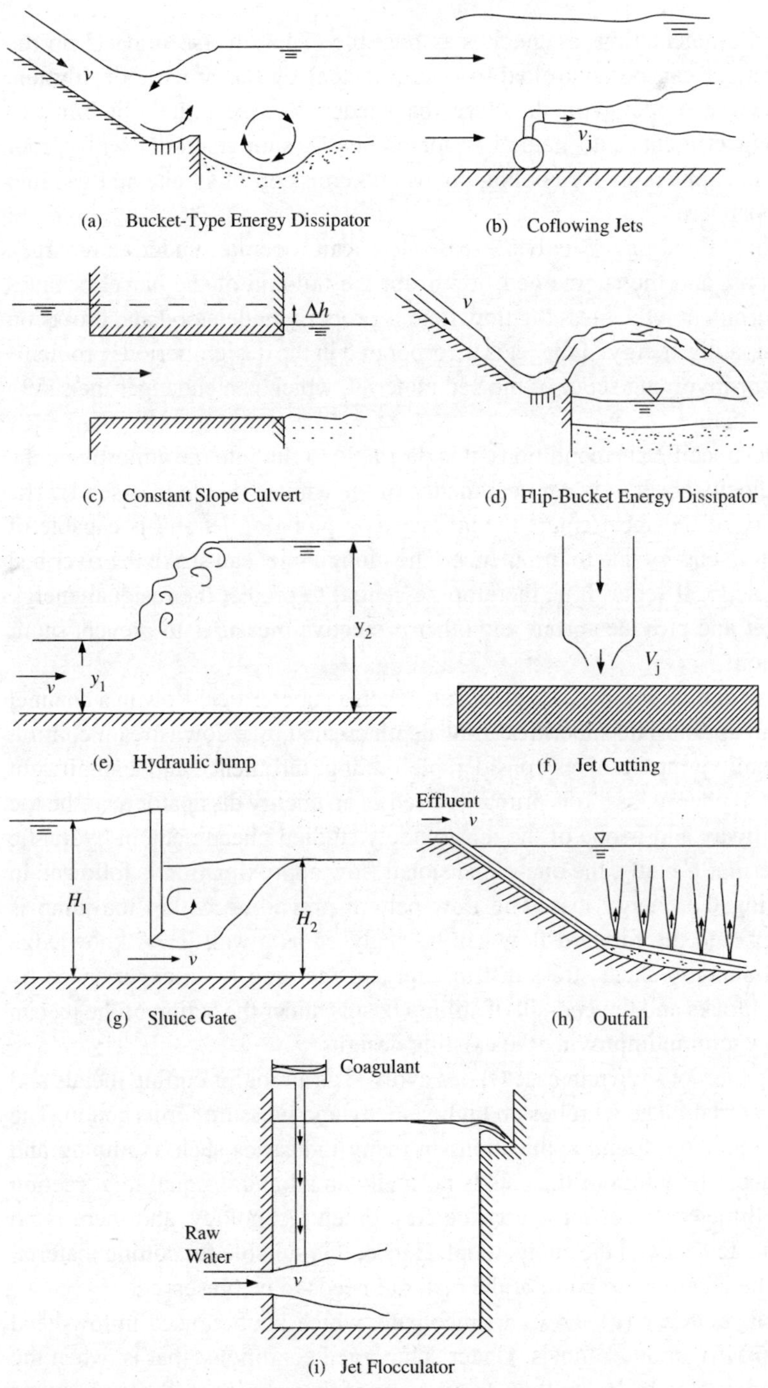

Figure 1.1 Typical examples of jets

with the ambient flow as quickly as possible. The adverse impacts on the environment can be controlled to a large extent by the process of dilution. A predictive model will, therefore, be needed to assess the suitability of disposing effluents into natural water bodies. A similar model with minor changes is applicable for cooling ponds and emission of smoke and gas into the atmosphere.

During flooding, a culvert, case (c), can operate under surcharged conditions, and the submerged flow from the tail-end of the barrel behaves like a turbulent jet. Unless the flow field is properly understood and provision for dissipating energy of the jet is incorporated in the design, serious problems can occur involving scour of the bed material, which can endanger the safety of the structure.

Under specific site conditions, it is desirable to flip into the atmosphere the high-velocity sheet of water near the toe of the dam as shown in case (d). The trajectory of the sheet enters the tailrace as a plunging jet and is capable of dissipating energy due to turbulence. The plunging jet can strike the river bed and cause local scour. It is, therefore, essential to predict the residual energy of the jet and provide aprons and other protective measures to prevent scour or erosion.

The hydraulic jump, case (e), occurs when a supercritical flow in a channel meets an appropriate subcritical flow depth created by a downstream control. A hydraulic jump involves considerable mixing, turbulence and a significant amount of energy loss. It is primarily used as an energy dissipator near the toe of a spillway and is one of the most widely studied phenomena in hydraulic engineering. Usually, the one-dimensional flow approximation is followed in estimating the energy loss. The flow pattern prevailing within the jump is complex, and the forward flow can be analysed as a wall jet. A knowledge of the boundary shear stress distribution and the drag force experienced by friction blocks and the end-sill of stilling basins under the action of the jet can be very useful in improving the existing designs.

The water jet shown in case (f) has evolved as a tool for cutting metals and other materials. The jet relies on high velocity and pressure for its action. The cutting jets have found applications in many industries such as mining and aerospace. The width of the cuts is normally small and is equal to a fraction of a millimeter. Water jet is considered a green technology, and there is no heat-affected zone in the cut material. Hence, it is suitable for cutting material where the inherent structure of the material needs to be preserved.

The layout case (g) shows a sluice gate, which is often used in low-head dams and irrigation channels. Under submerged conditions, that is, when the tailwater level is high, the flow coming out of the opening will behave like a wall jet and can cause bed scour. The classical wall jet model helps to analyse the flow beneath the sluice gate and assess the potential for scour. A concrete apron of suitable length can be provided to prevent scour.

Many coastal cities around the world dispose their sewage into the sea after some land-based primary treatment. The treated effluent is discharged through multiport sea outfalls, case (h), with the purpose of mixing the effluent with the sea water. Because the effluent is lighter than the sea water, buoyancy forces play an important role. The flow is treated as a buoyant jet in assessing the initial dilution near the surface of the sea. Here again, predictive models can play an important role in forecasting the dilution field and the related environmental impact assessment.

Raw water arriving at water treatment plants often contains fine suspended particles. In order to accelerate the removal of the suspended particles, water needs to be coagulated and flocculated. Such operations create large aggregates which can settle easily. The mechanical method is often used to create agitation in the flocculation chamber of water treatment plants. A hydraulic device such as the jet flocculator can be used as an alternative. The jet flocculator is almost as efficient as the mechanically driven flocculator in removing the suspended particles from water. The jet flocculator is simple in operation and is cost-effective. However, design guidelines are not available, and a proper understanding of the mixing phenomenon of the jet and the coagulant can help evolve the necessary guidelines. Furthermore, by predicting the energy loss in the flocculation chamber, one can foresee how efficiently the flocculator will work.

An understanding of the theory related to diffusion of jets can also help predict and analyse the energy losses in many other practical instances. For example, the inflow into a sewer junction box is akin to a jet, and the energy losses have been predicted for a 180° alignment of the inlet-outlet pipes.

2 Boundary Layer Approach

2.1 Introduction

A jet in the simplest form occurs when a fluid discharges with a uniform initial velocity, U_0, into a similar fluid medium, which is either at rest or in motion at a constant velocity, $U_a (< U_0)$. The difference in U_a and U_0 results in a tangential separation surface, causing eddies to form and move in a disorderly fashion. This results in an exchange of matter across the streams, and a region of finite thickness known as the 'jet boundary layer' is formed. In the jet boundary layer, a continuous distribution of velocity, species concentration and temperature takes place. The static pressure is constant throughout the flow in the boundary layer. Another feature of the shear layer is the smallness of the transverse velocity component over any section of the jet compared with the longitudinal velocity, u.

In the past, free-turbulence flow attracted great interest since it is the simplest case of turbulence, least affected by the viscosity of the fluid and easier to model. Several theories of free turbulence, involving different levels of sophistication, are in vogue. However, the following discussions pertain to the few widely used, successful theories.

Prandtl's old theory of free turbulence dates back to 1925 and was used by Tollmien to solve three problems of jet mixing (Abramovich, 1963). Using a single experimental constant, he was able to bring agreement between theory and experimental results. Later, Schlichting extended Prandtl's theory to the case of a wake flow behind streamlined bodies. The theory was used to find the temperature variation along the axis of the jet, as well as its distribution over the lateral cross-sections. Though Prandtl's theory implied complete similarity between the velocity and temperature fields, experiments showed that this was really not true. The contradiction was overcome by Taylor based on his vorticity transfer theory (Abramovich, 1963). On the basis of his theory and using only one experimental constant, the velocity and temperature profiles were found to agree with experiments. In spite of the fact that Taylor's theory did not give more accurate velocity profiles than Prandtl's theory, it has an advantage in being built on a more justifiable assumption. Hence, it is preferred in the cases involving temperature profiles and heat transmission.

A new theory of free turbulence was proposed by Prandtl in the year 1942 (Abramovich, 1963). He assumed the coefficient of turbulent viscosity to be proportional to the product of the density, the thickness of the jet and the velocity difference between the centreline and edge of the jet. The turbulent shearing stress was determined from Boussinesq's hypothesis. Using the new theory, Götler solved once again two problems related to submerged turbulent jets. The solution led to asymptotic velocity profiles at the lateral cross sections of the jet.

Even the new theory could not overcome the limitation of the old theory, that is, it could not differentiate between temperature and velocity distributions. However, the distribution of the mean velocity in Götler's analytical solution is Gaussian in nature and is easy to handle mathematically. Therefore, it has enjoyed greater acceptance in literature, and in this book, the solution methodology pertaining to it will be covered in brief. The final results of Tollmien's solution are also included for purpose of comparison. Details of the steps involved in the alternate solutions can be found elsewhere (Rajaratnam, 1976).

2.2 Boundary Layer of a Jet

Let us consider a two-dimensional flow of an incompressible fluid as shown in Fig 2.1.

Using the equation of motion in the x direction, the equation of continuity, and by averaging with respect to time, the following equation can be written

$$\frac{\partial \bar{u}}{\partial t}+\frac{\partial \bar{u}^2}{\partial x}+\frac{\partial \overline{u'^2}}{\partial x}+\frac{\partial \overline{uv}}{\partial y}+\frac{\partial \overline{u'v'}}{\partial y}=\nu\frac{\partial^2 \bar{u}}{\partial y^2}-\frac{1}{\rho}\frac{\partial \bar{p}}{\partial x} \tag{2.1}$$

In the above equation, $\bar{u}$ and $\bar{v}$ are the time averaged velocity components in the x and y directions, respectively. The corresponding fluctuating components are given by u' and v'. Since the velocity fluctuations change slowly with distance x, one can drop $\frac{\partial \overline{u'^2}}{\partial x}$ from Eq 2.1.

It may be noted that the turbulent shear stress $\tau_{xy}=\tau_{yx}=-\rho\overline{u'v'}$, and that the mean velocities $\bar{u}$ and $\bar{v}$ satisfy the continuity equation just as the instantaneous velocities do. For simplicity, let us avoid writing the bars over the mean values of the velocities. Unless specified otherwise, u and v would signify the time-averaged mean velocity components. The equations of motion,

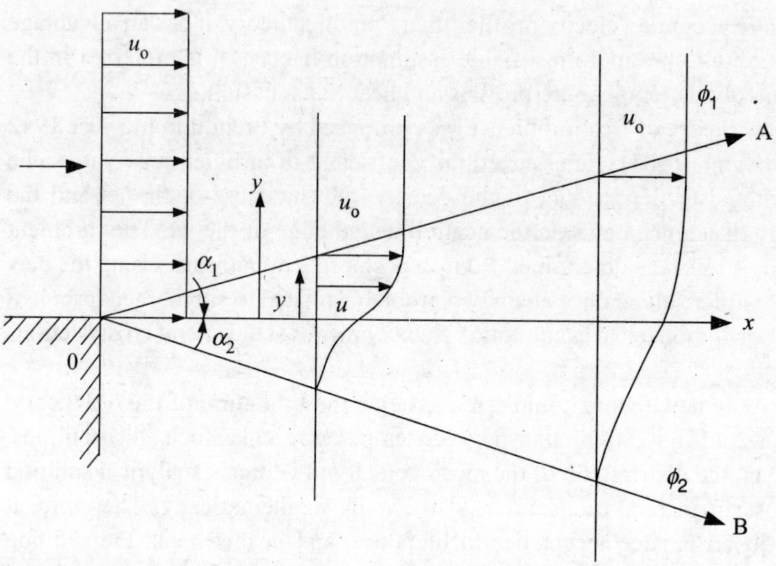

Figure 2.1 Definition sketch of a jet boundary layer

therefore, take the form

$$u\frac{\partial u}{\partial x}+v\frac{\partial u}{\partial y}=-\frac{1}{\rho}\frac{\partial \tau_{xy}}{\partial y} \tag{2.2}$$

and

$$\frac{\partial u}{\partial x}+\frac{\partial v}{\partial y}=0 \tag{2.3}$$

According to Prandtl's new theory, $\tau_{xy}=\rho\varepsilon\frac{\partial u}{\partial y}$, where the kinematic eddy viscosity of a submerged jet is

$$\varepsilon=\kappa\ b\ (u_m-U_a) \tag{2.4}$$

In Eq 2.4, u_m and U_a are the values of the flow velocity at the edges of the mixing zone at a given cross section, b is the width of the mixing zone and κ is an empirical constant.

It may be observed that in Eq 2.4, ε has a constant value over any given cross section even though it is a function of x. This is so because both b and (u_m-U_a) vary only with the streamwise distance, x. Equation 2.2 can be rewritten in the following form based on the shear-stress relationship of the new theory

$$u\frac{\partial u}{\partial x}+v\frac{\partial u}{\partial y}=\varepsilon\frac{\partial^2 u}{\partial y^2} \tag{2.5}$$

Table 2.1 Götler's theory

ξ	$\frac{u}{U_0}$	ξ	$\frac{u}{U_0}$
∞	1.0	−0.6	0.198
1.6	0.988	−0.8	0.129
1.4	0.976	−1.0	0.079
1.2	0.955	−1.2	0.045
1.0	0.921	−1.4	0.024
0.8	0.871	−1.6	0.012
0.6	0.802	−1.8	0.005
0.4	0.714	−2.0	0.002
0.2	0.613	−2.2	0.001
0.0	0.500	−2.4	0.00
−0.2	0.388	−2.6	0.00
−0.4	0.286	$-\infty$	0.0

Introducing certain change of variables and transformations, the above governing equation has been recast as an ordinary differential equation (Abramovich, 1963),

$$F''' + 2\sigma F F'' = 0 \tag{2.6}$$

Equation 2.6 was derived using the below mentioned relationships

$$\begin{aligned} &\frac{u}{U} = \sigma F'(\xi); \ \xi = \sigma\frac{y}{x}; \quad U = \frac{1}{2}(u_m + U_a) \\ &b = kx; \quad \psi = \int u\, dy = xUF(\xi); \quad \frac{v}{u} = -\frac{\partial \psi}{\partial x} = \xi F' - F \\ &\frac{1}{\sigma} = 2\sqrt{\kappa k \lambda}; \quad \lambda = \frac{u_m - U_a}{u_m + U_a} \end{aligned} \tag{2.7}$$

In the specific case of the boundary layer forming at the edge of a submerged jet, as shown in Fig 2.1, the requisite velocities are $u_m = U_0,\ U_a = 0$ *and* $U = U_0/2$. For such a flow, the solution of Eq 2.6 has been found as

$$\frac{u}{U_0} = 0.5 + \frac{1}{\sqrt{\pi}}\int_0^{\xi} e^{-z^2} dz \tag{2.8}$$

Using an error function integral table, the final answers from Götler's theory for the u and v components can be found in Table 2.1.

To bring agreement with Reichardt's experimental data, a value of $\sigma = 13.5$ was selected (Abramovich, 1963). Figure 2.2 shows the theoretical profiles

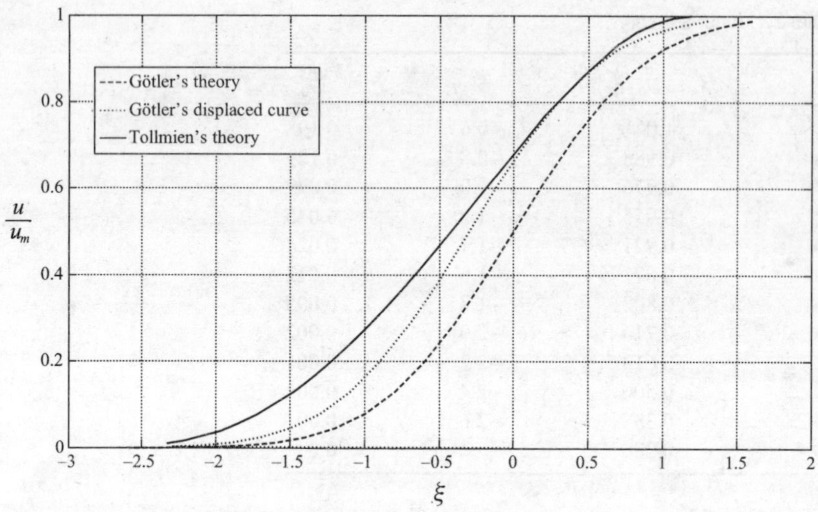

Figure 2.2 Velocity distribution in the jet boundary layer

obtained by Götler (new theory) and Tollmien (old theory). For a better match with the available experimental data points, the profile obtained from Eq 2.8 needs a displacement in the positive y direction. In other words, one has to assume that the nozzle edge passes along the line $\xi_0 = 0.3$ instead of $\xi_0 = 0$.

Thus, in the case of the boundary layer flow, Götler's theory requires two experimental constants, σ and ξ_0. In contrast, Tollmien's theory requires just one experimental constant 'a'. In case of turbulent sources such as a plane jet or an axially symmetric nozzle, a single experimental constant σ suffices to predict the velocity distributions which agree with observed values. The axially symmetric case will be dealt with in some detail in Chapter 3.

At this stage, it may be of interest to note that the velocity distribution in the jet boundary layer can also be found based on Reichard's hypothesis. This approach (presented later in Chapter 4) gives the distribution of the momentum flux as

$$\frac{u^2}{U_0^2} = \frac{1}{2}\left\{1 + erf\left(\frac{y}{\sqrt{2}b}\right)\right\} \tag{2.9}$$

where the characteristic length scale 'b' is a linear function of x and can be expressed as $b = Cx$, where C is the constant of proportionality.

The velocity distribution resulting from Eq 2.9 is given in Table 2.2, where $\zeta = \frac{y}{\sqrt{2}Cx}$.

Table 2.2 Point-Source method

ξ	$\frac{u}{U_0}$	ξ	$\frac{u}{U_0}$
2.0	0.999	−0.1	0.666
1.0	0.960	−0.3	0.579
0.70	0.916	−0.4	0.534
0.50	0.872	−0.5	0.489
0.40	0.845	−0.7	0.401
0.30	0.815	−1.0	0.280
0.10	0.746	−1.5	0.130
0.00	0.707	−2.0	0.048

In order to compare the analytical velocity distributions from the three theories, namely Tollmien's, Götler's and Reichardt's, it is necessary to know the relationship between the different coordinate systems adopted by them. The equivalence between ζ (Reichardt), ξ (Götler) and φ (Tollmien) can be expressed as

$$\zeta = \left(\frac{1}{\sigma\sqrt{2C}}\right)\xi = k_1\xi \tag{2.10}$$

and

$$\zeta = \left(\frac{a}{\sqrt{2C}}\right)\varphi = k_2\varphi \tag{2.11}$$

Selecting average experimental values of $C = 0.06$, $\sigma = 13.5$ and $a = 0.09$, the coefficients k_1 and k_2 work out as 0.871 and 1.059, respectively.

Albertson et al. (1950) have proposed an expression for the mean velocity distribution in the diffusion region of a two-dimensional flow-establishment zone. For a choice of the co-ordinate system shown in Fig 2.1, their expression can be simplified as

$$log_{10}\frac{u}{U_0} = -18.4\left(0.096 + \frac{y}{x}\right)^2 \tag{2.12}$$

While plotting the distribution given in Eq 2.12, the sign of y needs to be changed to tally with the system adopted for the jet boundary layer flow.

Figures 2.3 and 2.4 show that the differences in the velocity distribution from various theories namely Tollmien's, Reichardt's and Albertson's are quite small. The agreement between Götler's solution and Reichardt's method is not that good and is not shown here.

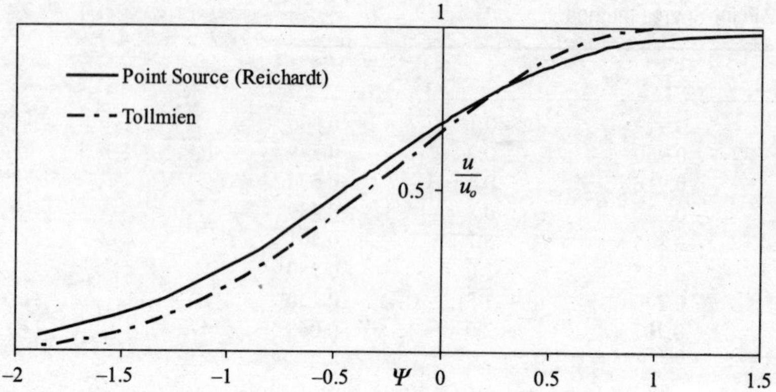

Figure 2.3 Comparison of Tollmien's and Reichardt's solutions

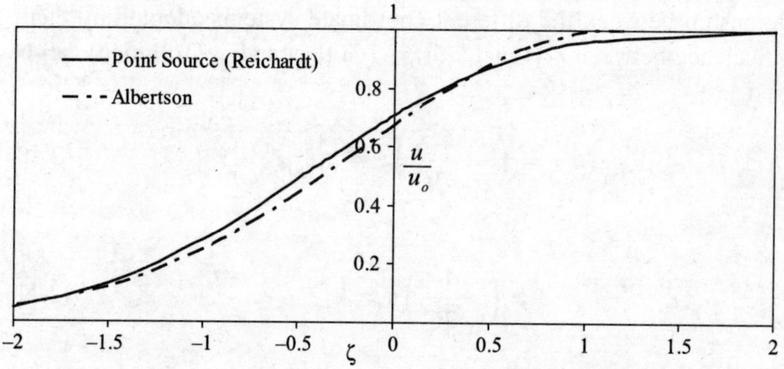

Figure 2.4 Comparison of Albertson's profile with Reichardt's solution

2.3 Summary

When two parallel streams of fluid having velocities U_0 and U_a meet, a shear layer appears whose thickness is small in the lateral direction. Therefore, a boundary layer type of approximation has been applied to simplify the equations of motion. The velocity distribution in the shear layer, that is, the jet boundary layer, was derived in the past by Tollmien and Götler based on Prandtl's old and new theories of free turbulence, respectively. It has been shown that one can also derive a velocity distribution based on Reichardt's momentum transfer theory, which agrees with Tollmien's solution. In all the theoretical approaches, experimental coefficients are needed for predicting the velocity distribution across the jet boundary layer.

Example 2.1: A thick stream of air, U_0(u_0 in the figure) = 2 m/s, enters a valley as shown in the figure.

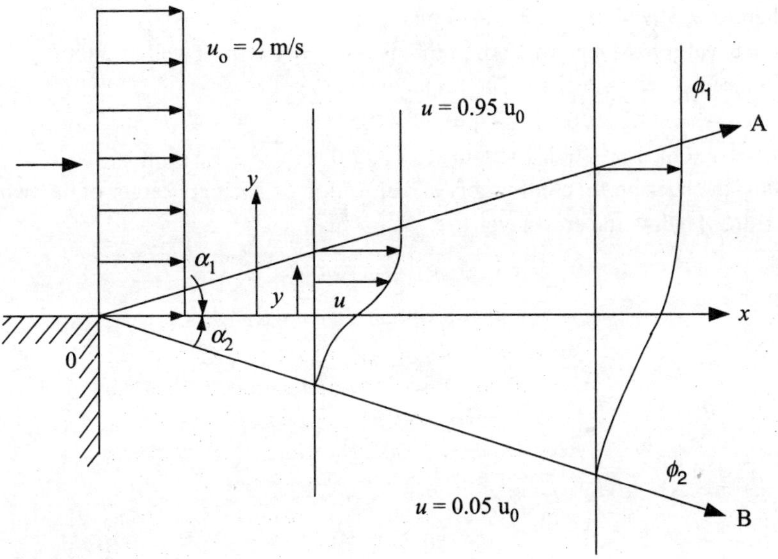

Example 2.1

Find

(a) The angles α_1 and α_2 which demarcate the shear layer 'b' that has the limiting values of $\frac{u}{U_0} = 0.95$ and 0.05, respectively.
(b) The forward velocity u for the $y = 0$ level at the distances $x = 1.0$ and 3.0 m.
(c) If a bumble bee can fly comfortably against an air current of no more than 0.3 m/s, what should be the maximum elevation of the plane in which it should hover in the valley? It maintains a minimum distance of 2.0 m and a maximum distance of 12.0 m from the vertical face of the cliff.

Assume $C = 0.06$ in Reichardt's theory.

Solution:

(a) In Reichardt's theory, $\zeta = \frac{y}{\sqrt{2Cx}} = \frac{y}{(0.085)x}$
From Table 2.2, corresponding to $\frac{u}{U_0} = 0.95$, $\zeta \simeq 1.0$. Therefore, $\frac{y}{x} = 0.085 \times (1.0) = 0.085$.
$\alpha_1 = \tan^{-1}(0.085) = 4.86°$
Similarly, corresponding to $\frac{u}{U_0} = 0.05$, $\zeta \simeq -2.0$. Therefore, $\frac{y}{x} = -2.0 \times (0.085) = -0.0170$

Hence, $\alpha_2 = \tan^{-1}(-0.170) = -9.65°$.
The width $b = k \times x = (0.085 + 0.170)x = 0.255x$.

(b) For $y = 0$, $\zeta = 0$. From Table 2.2, the corresponding value of $\frac{u}{U_0} = 0.707$.
Therefore, $u = 0.707 \times 2 = 1.414$ m/s.
This is valid for every value of x, that is, for $x = 1.0$ m as well as 3.0 m.

(c) Required $\frac{u}{U_0} = \frac{0.3}{2.0} = 0.15$. The respective value of $\zeta = -1.3$.
Hence, $y = -1.3 \times (0.085) \times x = -0.1105 \times x$.
At $x = 2.0$ m, $y = -0.221$ m and at $x = 12.0$ m, $y = -1.326$ m.
Thus, the bee can fly comfortably at a elevation of 1.326 m (larger of the two y values) below the crest level.

3 Free Jets

3.1 Introduction

A free jet occurs due to a continuous discharge of a fluid from a source into a large body of the same or similar fluid. Typical example of a free jet is shown in Fig 3.1. The free jet helps to mix the fluid emanating from the source with the ambient fluid. The jet is termed as a free turbulent jet since it is not confined by solid boundaries, and the diffusion process is not affected by the molecular viscosity of the fluid. The jet Reynolds number for a circular jet is defined as $R_e = \frac{U_0 d}{\nu}$, where U_0 = efflux velocity, d = jet diameter and ν = kinematic viscosity of the fluid. In case of a plane jet, the characteristic dimension $2b_0$ replaces d in the definition of the Reynolds number. In general, if R_e exceeds 2000, jets are classified as turbulent jets. Turbulent flows, described in Chapter 1, are more common than laminar flows.

In free turbulence, the molecular stresses in the entire flow field are insignificant compared with the turbulent shear stresses. Hence, the analysis of free turbulent jets is generally easier than that of turbulent wall jets, where the viscous shear stress in the laminar wall region can not be ignored. The free turbulent jet, shown in Figure 3.1, has an important feature common with the boundary layer motion. The lateral extent of the jet, or the shear layer, denoted by width b, is small compared with the longitudinal distance, x. Hence, the velocity gradient in the lateral direction is much larger than in the x-direction.

The above features are in consonance with the assumption made by Prandtl to simplify the equations of motion in a boundary layer. Thus, the governing equations of motion for steady, incompressible jets are identical to the boundary layer equations. As a first approximation, it is reasonable to assume the pressure to remain constant, since the ambient pressure is 'impressed' on the jet. In the analysis followed herein, the pressure gradient $\frac{\partial p}{\partial x}$ will be considered as zero.

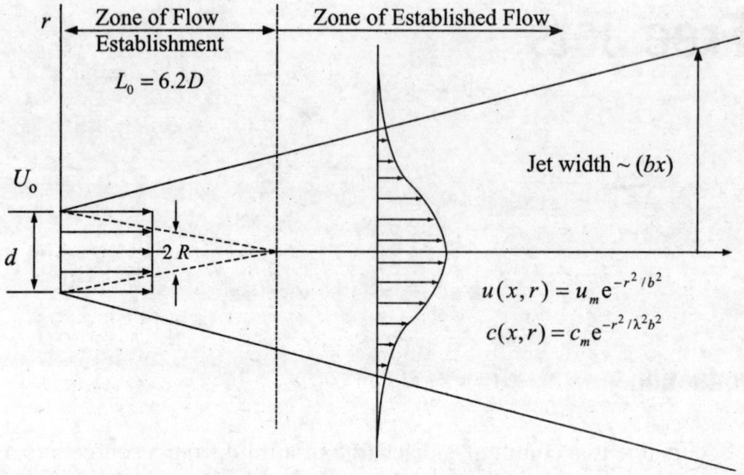

Figure 3.1 Axially symmetric jet: Definition sketch

3.2 The Axially Symmetric Jet

Figure 3.1 shows a jet of fluid discharging from a circular nozzle into a large expanse of fluid. The properties of the ambient fluid are the same as that of the jet fluid. The surrounding fluid gets entrained and mixes with the jet. This process of mixing is caused by the presence of large eddies that extend across the full width of the jet. Near the edge of the jet, the ambient fluid approaches the jet normal to the axis of the jet. As a consequence of entrainment, the volume flux, Q, past any section across the jet increases in the streamwise direction x.

Developing flow region

Experiments show that very close to the nozzle, the turbulence spreads inward and outwards from the edge of the jet. The inward spread from all around the edge forms a cone-like region of undiminished mean velocity U_0. This cone is popularly known as the potential core. It extends up to a distance of about 6.0 times the diameter of the outlet. It is followed by a very short transition zone, beyond which a region known as the fully developed flow region exists.

Fully developed flow region

In the fully developed flow region, the turbulence spreads over the entire thickness of the jet, and the mean velocities exhibit self-similarity. Therefore, by selecting suitable length and velocity scales, the velocity distribution at different streamwise sections can be represented by a single nondimensional distribution. The normalizing parameters which are in vogue are the length scale b and the velocity scale u_m. Though there is no ambiguity in the choice of u_m, the choice of the length scale, b, by different researchers is not unique. As it is extremely difficult and subjective to locate the distance of the edge of the jet from the centreline, b is commonly selected as that value of r where u equals some fraction of the maximum velocity, u_m. For example, the half-width of the jet is defined as that value of r at which the velocity u is $u_m/2$. The significance of other choices for b will be discussed later.

Self-similarity

A dimensionless plot of u/u_m verus y/b shows the distributions at different sections collapsing on to a single curve. This 'self-similar' property is clearly depicted in Fig 3.2. To take advantage of the principle of similarity of the velocity profiles, it will be necessary to predict the manner in which the length scale b and the velocity scale u_m vary with the distance x.

Several semiempirical approaches were adopted in the past to derive the similarity profile in the fully developed region of the jet. In many engineering applications, especially in water quality control, it suffices to assume the shape of the velocity distribution, a-priori. No matter how the velocity distribution is arrived at, an integral method comes very handy to derive analytically the mean flow characteristics. However, the constant appearing in the analytical solution needs to be arrived at from experiments. The integral approach is simple to use and does not require any assumption regarding the distribution of turbulence in the flow field.

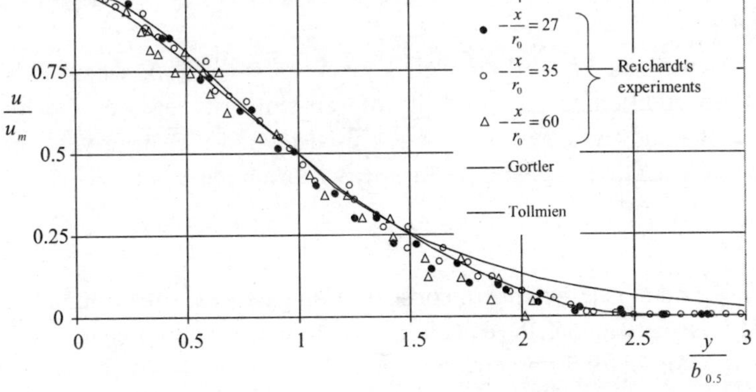

Figure 3.2 Similarity of mean velocity distribution across round jets (Source: Abramovich, 1963)

3.2.1 Equations of Motion

Ignoring the effects of molecular viscosity, for an incompressible round turbulent free jet, the time-averaged Reynolds equations of motion in cylindrical coordinates can be written as

$$\text{Continuity:} \qquad \frac{\partial u}{\partial x} + \frac{1}{r}\frac{\partial (rv)}{\partial r} = 0 \tag{3.1}$$

$$x\text{-Momentum:} \quad u\frac{\partial u}{\partial x} + v\frac{\partial u}{\partial r} = -\frac{1}{r}\frac{\partial (r\overline{u'v'})}{\partial r} = \frac{1}{\rho r}\frac{\partial (r\tau)}{\partial r} \tag{3.2}$$

where τ is the turbulent shear stress. The effect of molecular viscosity has been ignored. Let us also consider below the equation describing the mass conservation of tracer:

$$u\frac{\partial c}{\partial x} + v\frac{\partial c}{\partial r} = -\frac{1}{r}\frac{\partial (r\overline{u'c'})}{\partial r} \tag{3.3}$$

In the fully developed flow region, the velocity as well as the concentration profiles exhibit self-similarity. The distribution of the concentration across a round jet is depicted in Fig 3.3.

The velocity distributions can be found theoretically from the governing equations of motion making use of Prandtl's mixing length hypothesis (i.e., Tollmien solution) or by adopting the eddy viscosity model (Götler solution). Either of the models is essential since there are two equations but three unknowns u, v and τ. The closure model of Götler being simpler, it will be presented in brief.

3.2.2 Velocity Distribution

Götler solution

The eddy viscosity in Prandtl's model is given as $\varepsilon = \kappa u_m b$, where κ is a constant. In addition to the principle of self-similarity of the velocity distribution, the eddy viscosity model was used to find the mean velocity distribution of jets. Götler assumed the velocity distribution in a form

$$\frac{u}{u_m} = f(\xi) \tag{3.4}$$

where $\xi = \frac{\sigma r}{x}$ and σ is an unknown constant. Using Stokes' stream function and the eddy viscosity model, the equations of motion were simplified to arrive at the following ordinary differential equation (Rajaratnam, 1976).

$$FF' = F' - \xi F'' \tag{3.5}$$

The functions f and F are related as $f = \frac{F'}{\xi}$.

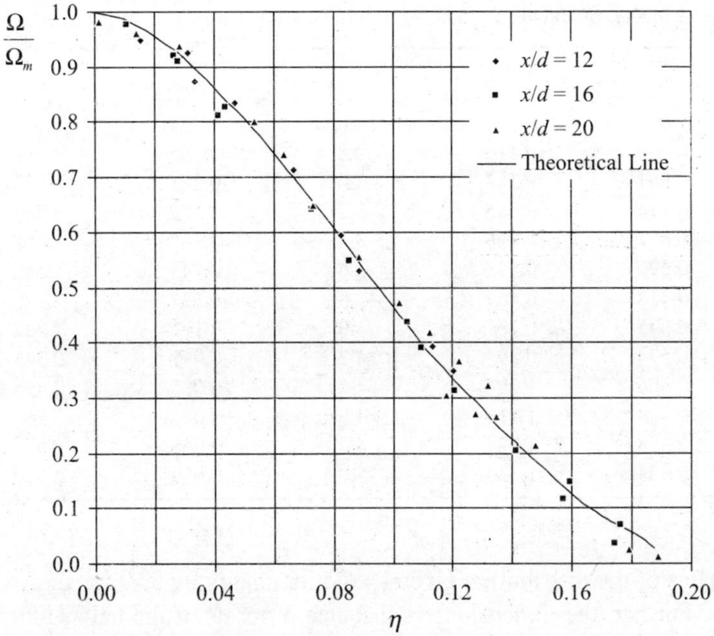

Figure 3.3 Similarity of concentration distribution for a round jet; Ω = concentration, $\eta = \frac{y}{x}$ (Source: Hinze and Zijnen, 1949)

Solution of the ordinary differential equation

It will be demonstrated that the velocity scale u_m is proportional to x^{-1}, whereas the length scale b grows linearly with the distance x. Consequently, for a circular jet the eddy viscosity, which is proportional to the product of u_m and b, attains a constant value throughout the flow field. Therefore, the solution of the turbulent jet will be identical to that of a laminar round jet, given by Schlichting (1968). The boundary conditions to be satisfied while solving Eq 3.5 are $u = u_m$ and $v = 0$ for $y = 0$. It means that $F' = 0$ and $F = 0$ for $\xi = 0$. The solution of the Eq 3.5 which satisfies these boundary conditions is given by

$$F = \frac{0.5\xi^2}{(1 + 0.125\xi^2)} \tag{3.6}$$

From the relationship $\frac{u}{u_m} = f = \frac{F'}{\xi}$, the velocity distribution is as follows:

$$\frac{u}{u_m} = \frac{1}{(1 + 0.125\xi^2)^2} \tag{3.7}$$

Table 3.1 Götler's theory: Round jet

$\xi = \frac{\sigma r}{x}$	$\frac{u}{u_m}$	$\eta = \frac{r}{b}$	$\xi = \frac{\sigma r}{x}$	$\frac{u}{u_m}$	$\eta = \frac{r}{b}$
0.0	1.00	0.0	2.6	0.294	1.450
0.2	0.990	0.116	2.8	0.255	1.562
0.4	0.961	0.223	3.0	0.221	1.673
0.6	0.915	0.335	3.2	0.192	1.785
0.8	0.857	0.446	3.4	0.167	1.897
1.0	0.790	0.558	3.6	0.145	2.008
1.2	0.718	0.669	3.8	0.127	2.120
1.4	0.645	0.781	4.0	0.111	2.232
1.6	0.574	0.893	4.2	0.097	2.343
1.8	0.506	1.004	4.4	0.085	2.455
2.0	0.444	1.116	4.6	0.075	2.566
2.2	0.388	1.227	4.8	0.066	2.678
2.4	0.338	1.339	5.0	0.059	2.789

The coordinates of the distribution given by Götler, that is Eq 3.7, are shown in Table 3.1. Further, the dimensionless distance in terms of the half-width, $\eta (= \frac{r}{b_{0.5}})$ is also shown as an additional column. It has been observed that Tollmien and Götler solutions give the velocity distribution with nearly the same accuracy, especially for $\eta \leq 1.5$. Unlike Götler's solution, Tollmien's solution gives a jet of finite thickness.

3.3 Gaussian Distribution

From experiments, it has been observed that the measured mean-velocity distribution across different streamwise locations follows a Gaussian distribution. Further, for all practical purposes, the differences between Tollmien's and Götler's solutions and the Gaussian distribution are relatively small, as shown in Fig 3.4. The semiempirical relationships for predicting the gross parameters of a jet, resulting from the integral type of analysis, are not sensitive to the nature of the velocity profile one assumes. Therefore, to take advantage of the simplicity in the mathematical operations, the Gaussian distribution of velocity is frequently used for the developed flow region of the jet. Typically, the Gaussian distribution is expressed in any of the following forms:

$$\frac{u}{u_m} = e^{-\left(\frac{r}{b_1}\right)^2} = e^{-\xi^2} \qquad \text{where} \quad \xi = \frac{r}{b_1} = \frac{r}{1.414\sigma} \tag{3.8}$$

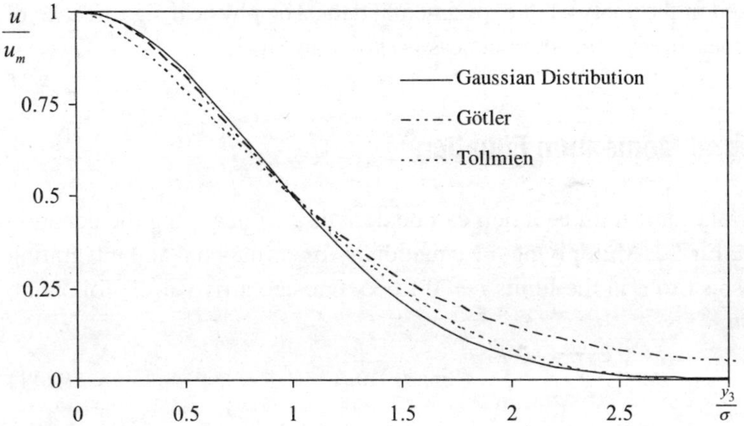

Figure 3.4 Comparison between different velocity distributions

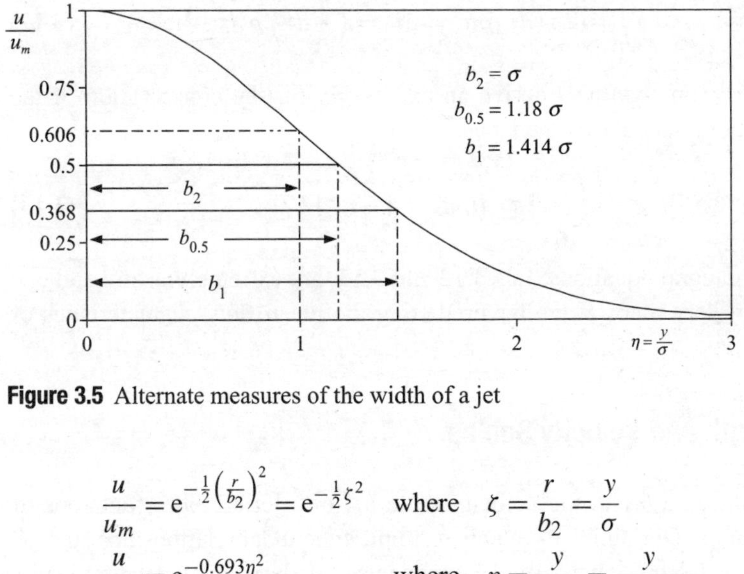

Figure 3.5 Alternate measures of the width of a jet

$$\frac{u}{u_m} = e^{-\frac{1}{2}\left(\frac{r}{b_2}\right)^2} = e^{-\frac{1}{2}\zeta^2} \quad \text{where} \quad \zeta = \frac{r}{b_2} = \frac{y}{\sigma} \tag{3.9}$$

$$\frac{u}{u_m} = e^{-0.693\eta^2} \quad \text{where} \quad \eta = \frac{y}{b_{0.5}} = \frac{y}{1.18\sigma} \tag{3.10}$$

In the above equations, σ is the standard deviation of the normal distribution curve.

It is important to note that the three different measures of the width of the jet have been expressed in terms of the standard deviation σ. Therefore, the measures can be easily converted from one to the other, while analyzing data from different sources. The half-width $b_{0.5}$ seems to be convenient and has been widely used in the analysis of experimental data. The physical significance of the three measures of the jet thickness is shown in Fig 3.5.

3.4 Integral Momentum Equation

The integral momentum equation can be derived by integrating the equation of motion, Eq 3.2. Multiplying the equation by ρr throughout and integrating it with respect to r in the limits $r = 0$ to ∞, one can arrive at the following expression.

$$\frac{d}{dx}\int_0^\infty (2\pi r dr)\rho u^2 = 0 \tag{3.11}$$

It shows that the momentum flux in the axial direction is conserved. Alternately, we can also express Eq 3.11 as

$$\int_0^\infty (2\pi r dr)\rho u^2 = M_0 = \left(\frac{\pi}{4}d^2\right)\rho U_0^2 \tag{3.12}$$

Similarly, one can also derive an expression for the conservation of the mass-flux of tracer across the jet as

$$\int_0^\infty (2\pi r dr)uc = \left(\frac{\pi}{4}d^2\right)U_0 c_0 \tag{3.13}$$

These integral equations, Eqs 3.12 and 3.13, are extremely useful and can be used in very many ways for predicting the mean flow characteristics of turbulent jets.

3.5 Length and Velocity Scales

The length scale b and the velocity scale u_m of a jet are both functions of the distance x. Our aim is to establish simple functional relationships such as $u_m \propto x^p$ and $b \propto x^q$, where p and q exponents are to be evaluated theoretically. To do so we need to set up two equations in p and q.

First equation in p and q

As the exact nature of the similarity function is not that very important, let us write it in a general form

$$\frac{u}{u_m} = f\left(\frac{r}{b}\right) = f(\eta) \tag{3.14}$$

Therefore, making use of Eq 3.14, the momentum conservation equation, Eq 3.11, can be written as

$$\frac{d}{dx}(\rho u_m^2 b^2)\int_0^\infty (2\pi\eta)f^2 d\eta = 0 \tag{3.15}$$

It may be noticed that the value of the integral in Eq 3.15 will depend upon the nature of the function $f(\eta)$. Nonetheless, it will always be a finite quantity.

Thus,

$$\frac{d}{dx}(u_m^2 b^2) = 0 \tag{3.16}$$

Therefore, Eq 3.16 leads to the following relationship between the exponents:

$$u_m^2 b^2 \propto x^0 \tag{3.17}$$

or

$$p + q = 0 \tag{3.18}$$

Second equation in p and q

To formulate a second equation, one can take recourse to many alternatives. Let us adopt the entertainment hypothesis proposed by Morton et al. (1956). The velocity with which the ambient fluid gets entrained into the jet, v_e, is proportional to the difference in velocity that exists between the jet centreline u_m and the ambient velocity u_a. For the stagnant ambient under consideration, $u_a = 0$. It was assumed by Morton that the direction of the entrainment velocity is normal to the axis of the jet. Hence, from the principle of conservation of mass, we can write

$$\frac{dQ}{dx} = \frac{d}{dx}\int_0^\infty (2\pi r dr)u = 2\pi \bar{b} v_e \tag{3.19}$$

In the above equation, $\bar{b}$ is the radius of a nominal outer boundary of the jet where u is negligible and v_e is normal to the x-axis. The exact location of the nominal boundary is of little consequence as the ratio between b and $\bar{b}$ is a fixed quantity, so long as the velocity profiles satisfy self-similarity.

Thus, the mass conservation equation can be rewritten as

$$\frac{d}{dx}(u_m b^2)\int_0^\infty 2\pi\eta f d\eta = 2\pi \bar{b} v_e \tag{3.20}$$

or

$$\frac{d}{dx}(u_m b^2) \propto \bar{b} v_e \propto b u_m \propto x^{p+q} \tag{3.21}$$

By equating the exponents of x on either side of Eq 3.21, we have $p+2q-1=p+q$. Therefore,

$$q=1 \tag{3.22}$$

Values of p and q

From Eqs 3.18 and 3.22, one can get the value $p=-1$. Thus, one could conclude that an axially symmetric jet grows with distance x in a linear fashion, while the maximum axial velocity decays in proportion to x^{-1}. Following a similar approach, for the case of a two-dimensional free jet it can be shown that $u_m \propto x^{-\frac{1}{2}}$ and $b \propto x$.

3.6 Decay of Maximum Velocity

The decay of the maximum velocity of a free jet can be worked out based on the momentum conservation equation. In the integral, one can assume a distribution for the velocity or else use the derived velocity distribution.

Decay based on Gaussian velocity distribution

For ease of analysis, the distribution of the velocity can be assumed to be Gaussian in nature, that is, $\frac{u}{u_m}=\mathrm{e}^{-\frac{r^2}{b^2}}$. By substituting for u $(=u_m\mathrm{e}^{-\frac{r^2}{b^2}})$ in the momentum conservation equation, Eq 3.15, we have

$$M=2\pi\rho b^2u_m^2\int_0^\infty\left(\frac{r}{b}\right)\mathrm{e}^{-2\frac{r^2}{b^2}}d\left(\frac{r}{b}\right)=\frac{\pi}{2}(\rho b^2u_m^2) \tag{3.23}$$

Furthermore, at the efflux section

$$M_0=\rho U_0^2\left(\frac{\pi d^2}{4}\right) \tag{3.24}$$

As the momentum is conserved, we can equate the above two equations and write

$$\frac{u_m}{U_0}=\frac{1}{\sqrt{2}}\frac{1}{\frac{b}{d}}=\frac{1}{\sqrt{2}C}\frac{1}{\left(\frac{x}{d}\right)} \tag{3.25}$$

where C is the spreading coefficient in the relationship $b=Cx$.

Similarly, making use of the conservation of the flux of tracer at different sections along the path of the jet, an expression for the variation of the maximum tracer concentration can be derived as

$$\frac{c_m}{c_0}=\left[\frac{1+\lambda^2}{2\sqrt{2}\lambda^2}\right]\frac{1}{C}\frac{1}{\left(\frac{x}{d}\right)} \tag{3.26}$$

where λ is the spread ratio in the concentration distribution $\frac{c}{c_m}=\mathrm{e}^{-\frac{r^2}{\lambda^2b^2}}$

Decay based on Götler and Tollmien velocity distribution

If one uses the tabulated values of Götler's solution, Table 3.1, the integral appearing in the momentum integral equation, Eq 3.15, can be evaluated numerically. It will result in the following relationship.

$$\frac{u_m}{U_0} = \left(\frac{\sigma}{3.22}\right)\frac{1}{\left(\frac{x}{d}\right)} \tag{3.27}$$

A similar computation basd on Tollmien's solution (for details see Rajaratnam, 1976) leads to

$$\frac{u_m}{U_0} = \left(\frac{0.482}{a}\right)\frac{1}{\left(\frac{x}{d}\right)} \tag{3.28}$$

In summary, the velocity decay can be expressed as

$$\frac{u_m}{U_0} = \frac{C_u}{\frac{x}{d}} \tag{3.29}$$

where the numerical value of the constant C_u can be evaluated from any of the following:

$$\begin{aligned} C_u &= \frac{1}{\sqrt{2}C} && \text{In } Eq\ 3.25 \\ C_u &= \left(\frac{\sigma}{3.22}\right) && \text{In } Eq\ 3.27 \\ C_u &= \left(\frac{0.482}{a}\right) && \text{In } Eq\ 3.28 \end{aligned} \tag{3.30}$$

In all the above cases, the coefficients 'C' (Gaussian distribution), 'σ' (Götler's solution) or 'a' (Tollmien's solution) will have to be determined from experiments.

3.7 Dimensional Analysis

After traversing a short distance from the nozzle, the jet loses memory regarding the details of the initial conditions that exist at the efflux section. What is important for the dynamic behavior of a jet is its initial momentum flux M_0, which is preserved for the zero pressure gradient flow. Hence, it would seem reasonable to write the following functional relationship

$$u_m = \phi_1(M_0, \rho, x) \tag{3.31}$$

and

$$b = \phi_2(M_0, \rho, x) \tag{3.32}$$

The dimensionless equivalent form of Eq 3.31 is

$$\frac{u_m}{\sqrt{\frac{M_0}{\rho x^2}}} = C_u \tag{3.33}$$

As the momentum flux $M_0 = \rho U_0^2(\frac{\pi d^2}{4})$, Eq 3.33 can be written as

$$\frac{u_m}{U_0} = \frac{C_u}{\left(\frac{x}{d}\right)} \tag{3.34}$$

The velocity decay coefficient C_u can be determined from experiments.

Also, from Eq 3.32, the length scale can be written as $b = Cx$. Thus,

$$\frac{b}{d} = C\left(\frac{x}{d}\right) \tag{3.35}$$

In the above analysis, x is the distance measured from the virtual origin, a fictitious point, from which the momentum flux M_0 is imagined to emerge.

3.8 Experimental Results

Abramovich (1963) has presented the work of Trupel (1915), Reichardt (1942), Corrsin (1946), Hinze and Zijnen (1949), Albertson et al. (1950) and many others. However, in the present analysis, the work of Albertson will be used together with more recent data from a number of other sources.

Velocity decay

Experiments of Albertson resulted in a value of $C_1 = 0.114$, in the relationship $b_1 = C_1 x$, Eq 3.8, while the spread coefficient $\lambda = 1.12$. Subsequent measurements by others have yielded somewhat different values for C_1 and λ. Many of them are in the vicinity of $C_1 = 0.11$ and $\lambda = 1.16$. It is also of interest to note that the corresponding value of the commonly used half-width $b_{0.5} = 0.095x$.

Substituting for C in Eq 3.30, the following expression for the decay of the maximum velocity can be derived from Eq 3.29

$$\frac{u_m}{U_0} = \frac{1}{\sqrt{2}(0.114)}\frac{1}{\left(\frac{x}{d}\right)} = \frac{6.2}{\left(\frac{x}{d}\right)} \tag{3.36}$$

Therefore, the length of the potential core can be considered as $6.2d$.

Decay of maximum tracer concentration

Similar to the above approach, from Eq 3.26, the expression for the variation of the maximum tracer concentration corresponding to a value of $\lambda = 1.16$ can be written as

$$\frac{c_m}{c_0} = \frac{5.4}{\left(\frac{x}{d}\right)} \tag{3.37}$$

If it is assumed $\lambda = 1.2$, a minor change in the value of the numerator on the right hand side of Eq 3.37 occurs and it is equal to 5.26.

The other equation of practical interest, valid for $x \geq 6.2d$, is

$$\text{centreline dilution } S = \frac{c_0}{c_m} = \frac{1}{5.4}\left(\frac{x}{d}\right) = 0.185\left(\frac{x}{d}\right) \tag{3.38}$$

For a Gaussian distribution, the ratio of the centre-line concentration and the average concentration is $\simeq 1.73$. Hence,

$$\text{average dilution } \bar{S} = 0.32\left(\frac{x}{d}\right) \tag{3.39}$$

Dilution in terms of Q_0 and M_0

Often, it is convenient to write the average dilution in terms of the source volume flux and specific momentum flux. Since $d = \frac{2}{\sqrt{\pi}} \frac{Q_0}{M_0^{\frac{1}{2}}}$, from the above expression, the average dilution can be rewritten as

$$\bar{S} = \frac{0.28 M_0^{\frac{1}{2}} x}{Q_0} \tag{3.40}$$

From Eq 3.40, it is clear that for a given volume flux Q_0, one can attain a larger dilution along the path of the jet if M_0 can be made as large as possible, that is, by having a small diameter and a high flow velocity.

Volume and energy flux

As the average dilution is also equal to $\frac{Q}{Q_0}$, from Eq 3.40 we can express the volume flux Q at any specified distance from the outlet as

$$Q = 0.28 M_0^{\frac{1}{2}} x \tag{3.41}$$

The kinetic energy flux at any section is given as $E \sim \rho u_{\mathrm{m}}^3 b^2 \sim x^{-1}$. Based on the data of Albertson et al. (1950), the variation of the kinetic energy of the

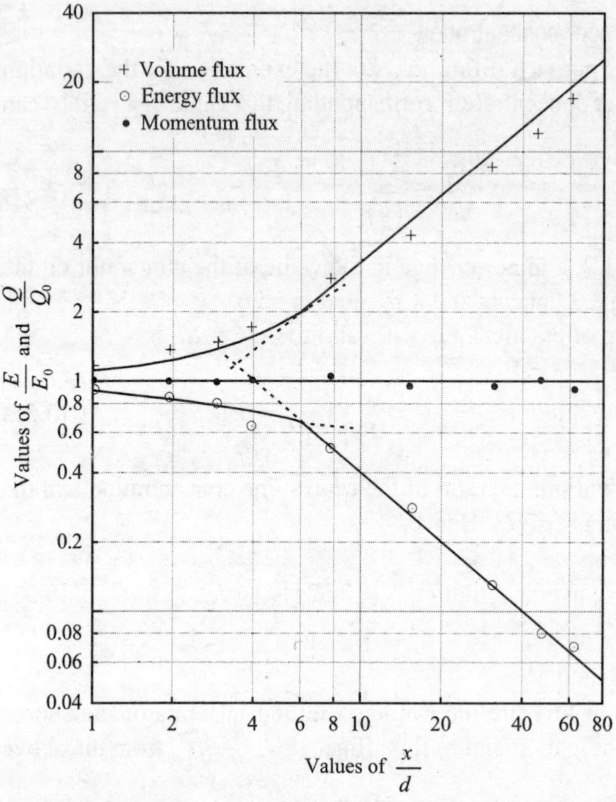

Figure 3.6 Downstream variation of volume and energy flux for a circular jet (Source: Albertson et al., 1950)

jet along the axis of the outlet can be shown as

$$\frac{E}{E_0} = \frac{4.1}{\left(\frac{x}{d}\right)} \tag{3.42}$$

The variations of $\frac{Q}{Q_0}$ and $\frac{E}{E_0}$ with the distance $\frac{x}{d}$ are shown in Fig 3.6. Good agreement between theory and experimental values can be observed in the figure.

Contrasting past and recent data

To validate the past classical data of Albertson, let us examine some of the recent data on velocity and concentration distribution observed by Chu (1996). The velocity measurements were made by laser doppler anemometry for a water jet having a diameter of $d = 10$ mm and a range of discharge velocities,

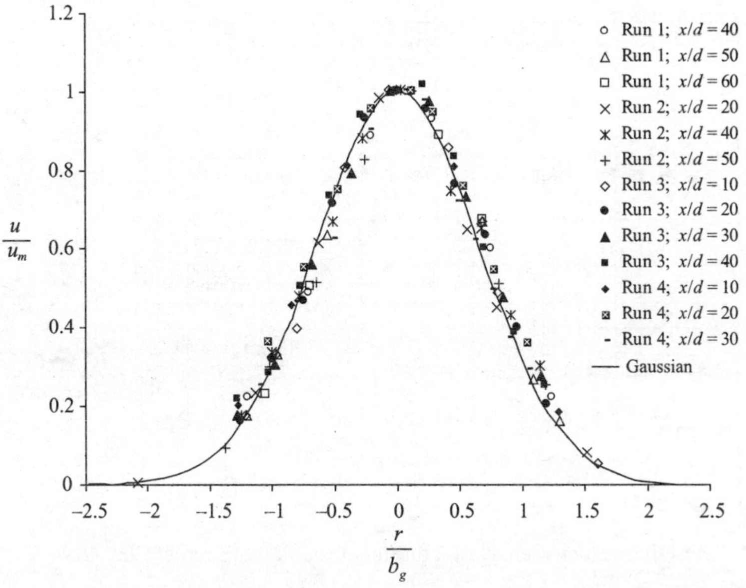

Figure 3.7 Radial profile of mean velocity in a round jet (Source: Chu, 1999)

$U_0 = 1.0, 0.9, 0.8, 0.7$ m/s. The nondimensional plot of $\frac{u}{u_m}$ against $\frac{r}{b_1}$ is shown in Fig 3.7. The velocity profiles are all self-similar and the distribution can be well-approximated by a Gaussian distribution. In Fig 3.7, b_g is the same as b_1 of Eq 3.8.

The measured data are also used in Fig 3.8 for showing the decay of the maximum velocity u_m with distance $\frac{x}{d}$. The agreement between the theoretical relationship, Eq 3.36, the recent data and the data of Albertson is noteworthy.

The recent measurements of the tracer concentration by Chu (1996), using the laser-induced fluorescence technique, are shown in Fig 3.9. The length scale for concentration distribution is taken as $b_c = \lambda b_1$. Here again, we observe that the profiles are self-similar and can be described quite well by a Gaussian distribution.

So far, we have discussed the behaviour of the time-averaged mean flow parameters as given by the fore-mentioned equations for velocity decay and variation of the maximum concentration. Even an elementary knowledge about the instantaneous velocity and concentration, coupled with important turbulence properties of the flow, can reveal interesting aspects about the entrainment mechanism and the ensuing dilution that occurs. Boundary layers bounded by a free stream, having negligible turbulence, are known to have

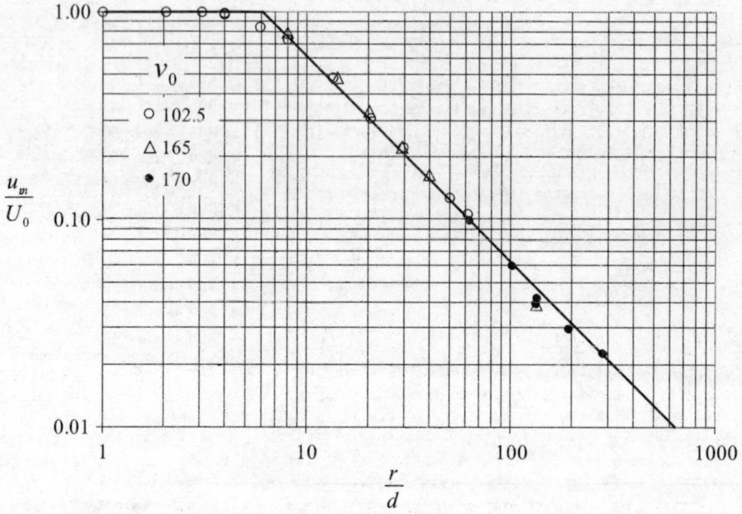

Figure 3.8 Centreline velocity decay of a turbulent round jet (Source: Chu, 1996)

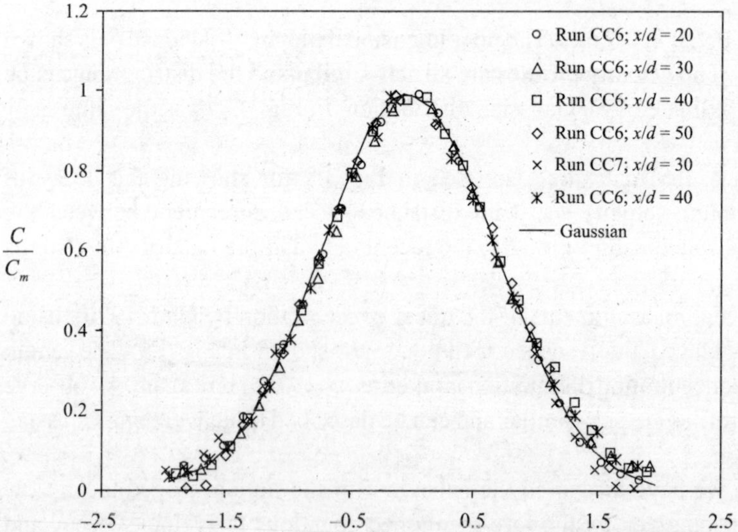

Figure 3.9 Observed tracer concentration distribution in a round jet (Source: Chu, 1996)

Figure 3.10 Irregular boundary of a round turbulent jet

sharp but irregular outer limits. This feature is common to all turbulent shear flows, including the jet, shown in Fig 3.10.

The boundary which separates the turbulent fluid from the nonturbulent ambient fluid is determined only by the extent to which the jet penetrates the surrounding. The large irregularities are primarily responsible for engulfing and entraining the surrounding fluid. If a probe, like the hot wire anemometer, is placed close to the edge of the jet boundary, the effects of the sharp, irregular boundary are discernible. The probe is surrounded by laminar and turbulent fluid lumps alternately, and the signal shows intermittent fluctuations. On processing such signals, the fraction of the time the flow is turbulent can be determined. This is defined as the intermittency factor γ. For a probe in the central core of the jet, the flow being turbulent all the time, the value of $\gamma = 1$. As we move out, close to the edge of the shear layer, the intermittency factor drops close to zero. This aspect is depicted in Fig 3.11 (Wygnanski and Fiedler, 1969).

The measured intensity of turbulence and the distribution of the turbulent shear stress across the jet are shown in Fig 3.12. The turbulent quantities exhibit self-similarity, and the maximum intensity of turbulence is nearly equal to 28–30 per cent.

The intensity of turbulent shear is zero on the axis of the jet because of symmetry and attains a maximum value near midway of the turbulent core. Another interesting feature is the distribution of the instantaneous maximum and minimum concentration in an axially symmetric jet. The measurements conducted by Chu (1996) are shown in Fig 3.13. The peak concentrations which occur near the centre of the jet can exceed the mean values by as much as 50 per cent.

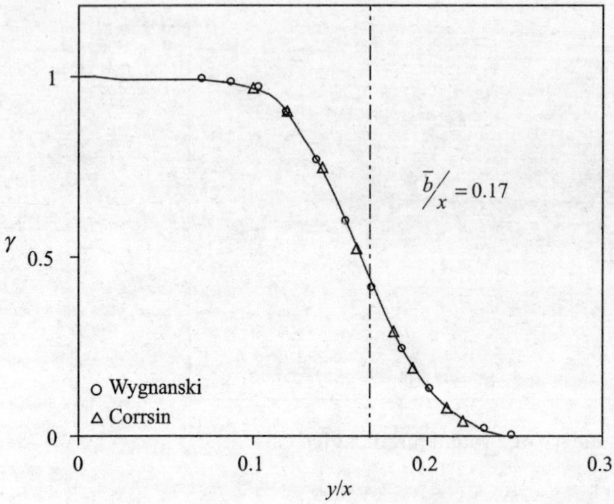

Figure 3.11 The intermittency-factor profile of a round jet (Source: Wygnanski and Fiedler, 1969)

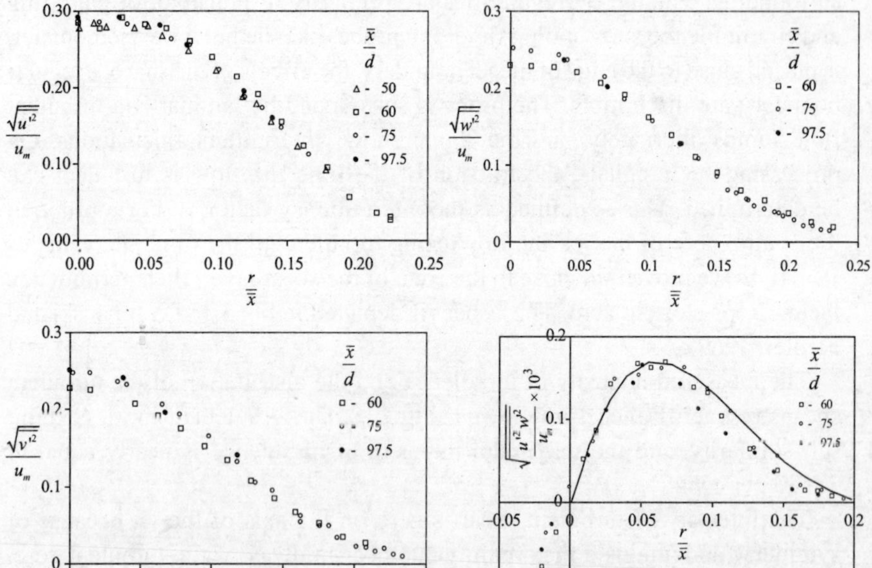

Figure 3.12 Observed intensity of turbulence and shear stress over the cross section of a circular jet (Source: Wygnanski and Fiedler, 1969)

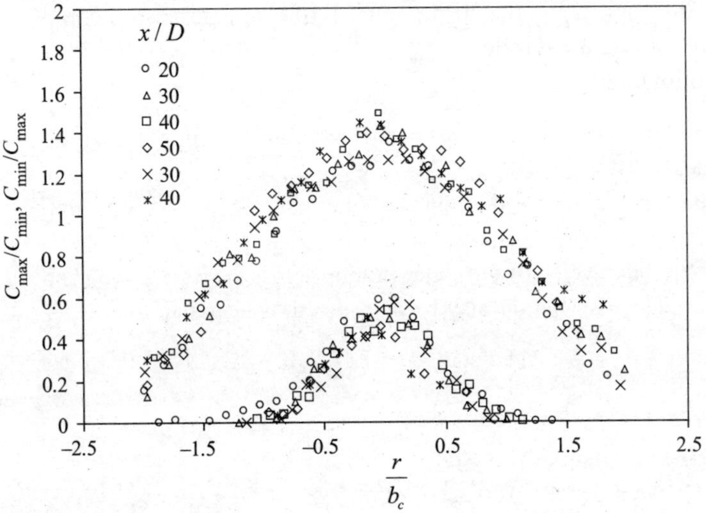

Figure 3.13 Radial profile of instantaneous maximum and minimum concentration in a circular jet (Source: Chu, 1996)

3.9 Summary

Important features of a free jet have been explained to lay the foundation for subsequent analysis. The equations of motion for an axially symmetric free jet have been simplified based on the boundary layer type of approximation. The flow field is in a developing state for $\frac{x}{d} \leq 6.2$ and can be considered as fully developed from thereon. The velocity and the tracer concentration exhibit self-similarity in the fully developed flow zone. The length scale $b \propto x$, whereas the velocity scale $u_m \propto x^{-1}$ for the axially symmetric jet.

Theoretical solutions of Tollmien and Götler are not significantly different from a Gaussian distribution. The axial momentum flux of the jet is conserved in case of zero pressure gradient flows.

The decay of the maximum velocity, the variation of the centreline tracer concentration and the dilution have been predicted from the integral equations. Experiments indicate that the maximum intensity of turbulence is about 28 per cent, and the instantaneous tracer concentration can vary by approximately ± 50 per cent from the mean values.

Example 3.1: A round air jet 100 mm in diameter, $\rho_{air} = 1.23\ \text{kg/m}^3$, is directed upward with a efflux velocity of $U_0 = 100$ m/s. A non-decaying tracer is injected into the supply pipeline at the rate of 100 mg/s to ensure complete mixing before leaving the nozzle.

Find the following:

(i) The maximum jet velocity and tracer concentration at a height of 2.0 m above the nozzle.

(ii) Centerline and average dilution at intervals of 1 m each, upto a height of 5 m.

(iii) The distribution of velocity and concentration across the jet at $x = 5$ m above the nozzle. The distribution may be considered as Gaussian, that is, $\frac{u}{u_m} = e^{-\frac{\xi^2}{2}}$, where $\xi = \frac{r}{0.085x}$ and assume $\lambda = 1.16$.

Solution:

$A = \frac{\pi}{4} \times (0.10)^2 = 7.854 \times 10^{-3}\ \text{m}^2$

$Q_0 = AU_0 = 7.854 \times 10^{-3} \times 100 = 0.7854\ \text{m}^3/\text{s}$

Mass flux of air $= \rho_{\text{air}} Q = 0.966$ kg/s

Mass flux of tracer $= 100 \times 10^{-6} = 10^{-4}$ kg/s

Tracer concentration, $c_0 = \frac{10^{-4}}{0.966} = 103.5$ ppm.

(i) $\frac{u_m}{U_0} = \frac{6.2}{\left(\frac{x}{d}\right)} = \frac{6.2}{(2/0.1)} = 0.31$ or $u_m = 0.31 \times 100 = 31.0$ m/s

Similarly, $\frac{c_m}{c_0} = \frac{5.4}{\left(\frac{x}{d}\right)} = \frac{5.4}{(2/0.01)} = 0.27$ or $c_m = 0.27 \times 103.5 = 27.9$ ppm.

(ii) The centreline dilution, $S = 0.185\left(\frac{x}{d}\right)$ and the average dilution, $\overline{S} = 0.32\left(\frac{x}{d}\right)$. Hence, the dilutions achieved at various locations along the axis of the jet are as follows:

x	$\frac{x}{d}$	S	$\overline{S}$
1	10	1.85	0.32
2	20	3.70	0.64
3	30	5.55	9.60
4	40	7.40	12.80
5	50	9.25	16.00

(iii) $\frac{u}{u_m} = e^{-\frac{\xi^2}{2}} = e^{-\frac{1}{2}\frac{r^2}{(0.085x)^2}} = e^{-69.2\left(\frac{r}{x}\right)^2}$

Similarly, $\frac{c}{c_m} = e^{-69.2\left(\frac{r}{\lambda x}\right)^2} = e^{-51.43\left(\frac{r}{x}\right)^2}$

The velocity and tracer concentration across the jet for the desired location, $x = 5$ m, is shown in the following table.

r, m	$\frac{r}{x}$	$\frac{u}{u_m}$	$\frac{c}{c_m}$	u (m/s)	c (ppm)
0.0	0.00	1.0	1.0	31.00	27.90
0.1	0.02	0.973	0.979	30.16	27.31
0.2	0.04	0.895	0.921	27.74	25.69
0.4	0.08	0.642	0.719	19.90	20.06
0.6	0.12	0.369	0.477	11.44	13.30
0.8	0.16	0.170	0.268	5.27	7.47
1.0	0.20	0.063	0.128	1.95	3.57

Example 3.2: For the conditions specified in Example 3.1, estimate the following quantities.

(i) The volume flux of air at a height of 5.0 m above the nozzle.
(ii) The efflux kinetic energy of the air jet and the loss in the kinetic energy by the time it reaches the height of 5.0 m.

If the dilution of the tracer is to be enhanced by a factor of 2 at the reference height $x = 5.0$ m, suggest means to achieve it.

Solution:

(i) The specific momentum, $M_0 = Q_0U_0 = 0.7854 \times 100 = 78.54\ \text{m}^4/\text{s}^2$

The volume flux, $Q = 0.28M_0^{\frac{1}{2}}x = 0.28 \times (78.54)^{\frac{1}{2}} \times 5 = 12.40\ \text{m}^3/\text{s}$

(ii) The energy at the nozzle, $E_0 = (\text{mass flux})\frac{u^2}{2} = (AU_0\rho)\frac{u^2}{2} = 0.7854 \times 1.23 \times \frac{100^2}{2} = 4830\frac{N-m}{s}$

$\frac{E}{E_0} = \frac{4.1}{(x/d)} = \frac{4.1}{(5/0.1)} = 0.082$; $E = 0.082 \times 4830 = 396.06\frac{N-m}{s}$

% Loss of energy $= \frac{(E_0 - E)}{E_0} = (1 - 0.082) \times 100 = 91.8$ per cent

The average dilution at a distance of x from the nozzle is $\overline{S} = \frac{0.28M_0^{\frac{1}{2}}x}{Q_0}$.

Thus, for given values of Q_0 and x, $\overline{S} \propto M_0^{\frac{1}{2}}$.

Or, $\frac{\overline{S}_1}{\overline{S}_2} = \frac{(Q_0U_0)_1^{\frac{1}{2}}}{(Q_0U_0)_2^{\frac{1}{2}}} = \sqrt{\frac{U_{01}}{U_{02}}}$

As the average dilution is to be increased by a factor of 2, the velocity ratio has to increase by a factor of 4. Hence, $U_{02} = 4U_{01} = 4 \times 100 = 400$ m/s.

From the relationship, $Q_0 = \frac{\pi}{4}d_2^2 \times U_{02}$, the diameter of the nozzle $d_2 = 50$ mm.

4 Reichardt's Momentum Transport Theory

4.1 Introduction

Similarity exists in the transmission of shear in molecular physics and the mechanics of turbulence. In the earlier chapters, the molecular concept of mean free path was adopted to represent the exchange of momentum between layers based on Prandtl's mixing length hypothesis. There are other theories as well which predict the mean velocity distribution. Establishing a link between the turbulent shearing stresses and the mean motion is the common feature of all the theories. The velocity distributions obtained from different turbulence models show marginal differences. In all the cases, a final validation with experiments is inevitable. They all belong to deductive theory of turbulence, and it is difficult to ascertain, *a-priori*, as to which of the alternatives is closest to physical reality.

Reichardt (1941) attempted to formulate an inductive theory of turbulence. He applied the theory of thermal conduction to the theory of turbulent propagation. The equation for molecular diffusion or heat conduction yields a Gaussian solution. A close examination of the then available data, supplemented by his own experiments, showed that the total momentum flux $\rho\overline{U^2}$, which includes the mean and turbulent components, follows the Gaussian error curve. Hence, the differential equation for $\rho\overline{U^2}$ must be identical with the equation for heat conduction. To transform the equations of motion to a form identical to the heat conduction equation, a new hypothesis was introduced. The essence of Reichardt's theory is elaborated in the following sections.

4.2 Theory of Turbulent Mixing

New theory of momentum transport

Reichardt assumed that the lateral transport of momentum is proportional to the transverse gradient of the horizontal component of momentum, that is, for a given density of the fluid

$$\overline{UV} = -\Lambda \frac{\partial \overline{U^2}}{\partial y} \tag{4.1}$$

In the case of an axially symmetric flow, y will be replaced by r.

Let us consider the case of an axially symmetric flow with its axis oriented in a horizontal direction. For simplicity, we shall ignore the horizontal pressure gradient and the viscous effects in the governing equation of motion. The continuity equation and the x-component of the momentum equation, in terms of the instantaneous velocities U and V are, respectively, as follows

$$\frac{\partial U}{\partial x} + \frac{1}{r}\frac{\partial (rV)}{\partial r} = 0 \tag{4.2}$$

$$U\frac{\partial U}{\partial x} + V\frac{\partial U}{\partial r} = 0 \tag{4.3}$$

Making use of the continuity equation in the momentum equation in the x-direction, Eq 4.3, can be recast in the following form

$$\frac{\partial \overline{U^2}}{\partial x} + \frac{1}{r}\frac{\partial (r\overline{UV})}{\partial r} = 0 \tag{4.4}$$

The over-bars denote averaging with respect to time. In consonance with experiments, let us further assume that the specific momentum-flux distributions in the lateral direction, at various stream-wise locations, are self-similar and can be expressed by

$$\frac{\overline{U^2}}{\overline{U_m^2}} = f(\xi), \tag{4.5}$$

where U_m is the instantaneous velocity on the axis of the jet, and the non-dimensional distance $\xi = \frac{r}{b}$. Based on Reichardt's momentum transport relationship, the equation of motion, Eq 4.4, can be modified as

$$\frac{\partial(\overline{U^2})}{\partial x} - \Lambda\left\{\frac{\partial^2(\overline{U^2})}{\partial r^2} + \frac{1}{r}\frac{\partial(\overline{U^2})}{\partial r}\right\} = 0 \tag{4.6}$$

Introducing the similarity relationship, Eq 4.5, in the above expression of Eq 4.6, the following can be written

$$2f + \xi f' = -\frac{\Lambda}{b}\frac{dx}{db}\left\{f'' + \frac{f'}{\xi}\right\} \tag{4.7}$$

The primes over the function f denote differentiation with respect to the variable ξ.

Solution for momentum-flux distribution

Since Eq 4.7 should only be a function of ξ and be independent of x, let us set $\frac{\Lambda}{b}\frac{dx}{db}$ = a constant. Hence, without loss of generality, we can set

$$\Lambda = b\frac{db}{dx} \tag{4.8}$$

As the characteristic width b is a function of x, it implies that Λ can only be a function of x. Thus, the final form of the governing equation of motion simplifies to

$$2f + f'\left(\xi + \frac{1}{\xi}\right) + f'' = 0 \tag{4.9}$$

The boundary conditions to be satisfied are, at $\xi = 0$, $f = 1$, and for $\xi \to \infty$, $f \to 0$. A solution of Eq 4.9 is

$$f = \frac{\overline{U^2}}{\overline{U_m^2}} = \mathrm{e}^{-\frac{\xi^2}{2}} \tag{4.10}$$

Decay of maximum momentum flux

Taking recourse to the conservation of the forward momentum flux, one can write

$$\int_0^\infty \rho\overline{U^2}(2\pi r dr) = \rho\overline{U_0^2}A \tag{4.11}$$

By substituting for $\overline{U^2}$ from Eq 4.10 in Eq 4.11, we can get

$$\frac{\overline{U_m^2}}{\overline{U_o^2}} = \frac{A}{2\pi b^2} \tag{4.12}$$

As the instantaneous velocity can be expressed as the sum of a time-averaged mean quantity u and the fluctuating component u', that is, $U = u + u'$, the time-averaged instantaneous flux of the momentum can be written as

$$\overline{U^2} = u^2 + \overline{u'^2} \tag{4.13}$$

Normally, $\overline{u'^2}$ is smaller than u^2 over most parts of the cross section of the round jet. Hence, as an approximation, it looks reasonable to drop the contribution of the fluctuating component in Eq 4.13. Thus, in terms of the mean velocities, the lateral distribution of the flux of the momentum and the variation of the square of the maximum mean velocity along the axis of the jet can be written as

$$\frac{u^2}{u_m^2} = \mathrm{e}^{-\frac{\xi^2}{2}} = \mathrm{e}^{-\frac{1}{2}\left(\frac{r}{b}\right)^2} \tag{4.14}$$

and

$$\frac{u_m^2}{U_0^2} = \frac{A}{2\pi b^2} \tag{4.15}$$

The point-source

If the physical size of the source is shrunk to an elemental area, dA, Eq 4.15 still being valid, can be rewritten as

$$\frac{u_m^2}{U_0^2} = \frac{dA}{2\pi b^2} \tag{4.16}$$

The difference between Eqs 4.15 and 4.16 is very subtle. However, it will be seen later that the form of the expression given in Eq 4.16 is very useful in predicting the velocity profiles for any general shape of an outlet. A source of momentum flux which satisfies Eq 4.16 will be known as the point-source. It serves as the basic 'building block' in the derivation of many formulae discussed in this book.

Advantages of the new momentum transport hypothesis

Though Reichardt's momentum transfer law helps to find the distribution of the momentum flux across a jet in a simple way, it is subject to some criticisms (Hinze, 1975). The theory does not conform with Newton's relativity which requires the forces in a mechanical system to be independent of the addition of a constant velocity. Like Prandtl's theories, the new momentum transfer law is not able to shed any light on the turbulent motion per se.

There is a strong case, nonetheless, in favour of Reichardt's momentum transfer law as it is capable of linearising the governing equation of motion involving $\overline{U^2}$. Hence, it has a very big advantage, not only mathematically but also from a practical point of view. The method of superposition of the elementary solutions can be adopted to solve for the flow from different shapes of outlets. Reichardt's theory can also be extended to cases in which the transport of a scalar quantity is involved, in lieu of the momentum flux component ρU^2

4.3 Distribution of Scalar Quantities

In this section, we will derive relationships for finding the distribution of temperature across a jet. The equations meant for temperatures are equally valid for the distribution of other scalar quantities such as the tracer concentration.

Governing equation for heat flux

For a heated jet, the absolute temperature T is the scalar quantity of interest. Using the continuity equation and by taking the time average of the transport equation for the scalar quantity, one can write.

$$\frac{\partial(\overline{UT})}{\partial x} + \frac{1}{r}\frac{\partial(r\overline{VT})}{\partial r} = 0 \tag{4.17}$$

It is convenient to work with temperature excess instead of absolute temperatures. Let us write $T = T_0 + \Delta T$, where the reference temperature T_0 does not change with x or r.

Hence, Eq 4.17 can be simplified into the following form:

$$\frac{\partial(\overline{U\Delta T})}{\partial x} + \frac{1}{r}\frac{\partial(r\overline{V\Delta T})}{\partial r} = 0 \tag{4.18}$$

Hypothesis for lateral transport of heat flux

Let us reintroduce Reichardt's hypothesis that the lateral transport of heat flux is proportional to the gradient of the heat flux in the radial direction, that is,

$$\overline{V\Delta T} = -\Lambda_t \frac{\partial(\overline{U\Delta T})}{\partial r} \tag{4.19}$$

A combination of Eqs 4.18 and 4.19 will lead to

$$\frac{\partial(\overline{U\Delta T})}{\partial x} - \Lambda_t \left\{ \frac{\partial^2(\overline{U\Delta T})}{\partial r^2} + \frac{1}{r}\frac{\partial(\overline{U\Delta T})}{\partial r} \right\} = 0 \tag{4.20}$$

Self-similarity of heat flux distribution

At this point, let us once again fall back on the similarity assumption regarding the distribution of heat flux by writing

$$\frac{\overline{U\Delta T}}{U_m \Delta T_m} = g(\zeta) \tag{4.21}$$

where $\zeta = \left(\frac{y}{b_t}\right)$, and b_t is a length scale that depends on x. From Eq 4.20 and 4.21, it can be shown that

$$2g + \left(\zeta + \frac{1}{\zeta}\right) g' + g'' = 0 \tag{4.22}$$

The boundary conditions to be satisfied are as follows: for $\zeta = 0$; $g = 1$, and for $\zeta \to \infty$; $g \to 0$.

Solution for heat flux

A solution of Eq 4.22 is

$$g = \frac{\overline{U\Delta T}}{\overline{U_m \Delta T_m}} = \mathrm{e}^{-\frac{\zeta^2}{2}}. \tag{4.23}$$

This is identical to the momentum flux distribution given in Eq 4.10. Since the heat flux is conserved, the following should hold good,

$$\int_0^\infty \rho U \Delta T (2\pi r dr) = \rho U_0 \Delta T_0 A \tag{4.24}$$

where ΔT_0 is the temperature excess in the plane of the nozzle.

Inserting the value of $U\Delta T$ from Eq 4.23 in the above expression, a final relationship works out as follows,

$$\frac{\overline{U_m \Delta T_m}}{\overline{U_0 \Delta T_0}} = \frac{A}{2\pi b_t^2} \tag{4.25}$$

where ΔT_m represents the excess temperature on the axis of the jet. Note that the length scale b_t is different than b.

Further, ignoring the contributions of the fluctuating quantities, the mean velocity and excess temperatures can be used in place of their instantaneous counterparts.

Till this point, the case of a round jet was discussed. The new hypothesis also leads to simple solutions for the jet boundary layer, as well as the plane slot. These cases are presented very briefly.

4.4 Two-dimensional Flows

For a steady two-dimensional turbulent flow, the continuity equation is

$$\frac{\partial U}{\partial x} + \frac{\partial V}{\partial y} = 0 \tag{4.26}$$

By ignoring the effects of the molecular viscosity, the momentum equation in x-direction can be written as

$$U\frac{\partial U}{\partial x} + V\frac{\partial U}{\partial y} = 0 \tag{4.27}$$

A combination of Eqs 4.26 and 4.27 results in

$$\frac{\partial U^2}{\partial x} + \frac{\partial UV}{\partial y} = 0 \tag{4.28}$$

Averaging Eq 4.28 over time, we get

$$\frac{\partial \overline{U^2}}{\partial x} + \frac{\partial \overline{UV}}{\partial y} = 0 \tag{4.29}$$

By adopting Reichardt's momentum transport relationship $\overline{UV} = -\Lambda(x)\frac{\partial \overline{U^2}}{\partial y}$, Eq 4.29 can be transformed to

$$\frac{\partial \overline{U^2}}{\partial x} = \Lambda \frac{\partial^2 \overline{U^2}}{\partial y^2} \tag{4.30}$$

Note that Eq 4.30 looks analogous to the heat conduction equation $\frac{\partial T}{\partial t} = a\frac{\partial^2 T}{\partial x^2}$. The boundary conditions are different in case of the jet boundary layer and the two-dimensional plane jet. Therefore, they will be dealt with separately and are presented below.

4.4.1 Jet Boundary Layer

The distribution of the instantaneous values of the momentum flux is assumed to be universal, as expressed in Eq 4.5. For a jet boundary layer, $U_m = U_0$ for all values of the distance x. Therefore, we can introduce the relationship $\overline{U^2} = \overline{U_0^2} f(\xi)$ into the governing equation for the jet boundary layer, Eq 4.30. On simplification, the governing equation reduces to

$$\xi f' = -\frac{\Lambda}{b}\frac{dx}{db} f'' \tag{4.31}$$

For the momentum flux profiles to be universal (independent of x-coordinate), $\frac{\Lambda}{b}\frac{dx}{db}$ should be a constant. Without loss of generality, let it be equal to unity. On integration of Eq 4.31,

$$f' = C_1 e^{-\frac{\xi^2}{2}} \tag{4.32}$$

A second time integration of the above expression results in

$$f = C_1 \int_0^{\xi} e^{-\frac{\xi^2}{2}} d\xi + C_2 \tag{4.33}$$

Constants of integration

For evaluating the two constants of integration C_1 and C_2, let us use the boundary conditions shown in Fig 4.1: for $\xi = 0$, $\overline{U^2} \approx 0.5\overline{U_0^2}$ (as observed from experimental data) and for $\xi \to \infty$, $f = 1$.

Thus, $C_2 = 0.5$ and $C_1 = \frac{0.5}{\int_0^\infty e^{-\frac{\xi^2}{2}} d\xi} = \frac{1}{\sqrt{2\pi}}$.

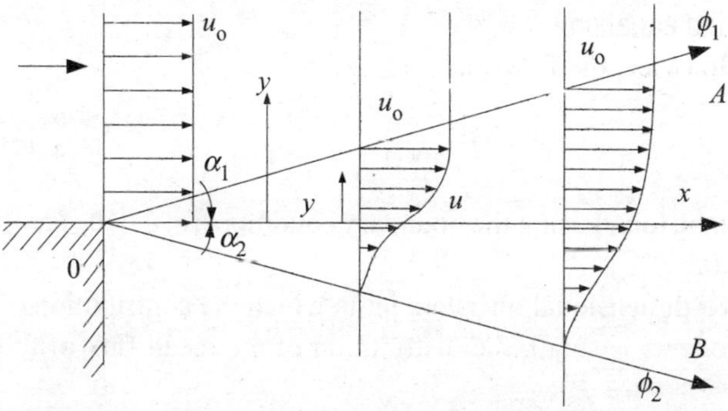

Figure 4.1 Definition sketch of a plane turbulent shear layer

Hence, for a two-dimensional jet boundary layer the distribution of the momentum flux in the shear layer is

$$\frac{\overline{U^2}}{\overline{U_0^2}} = 0.5 + \frac{1}{\sqrt{2\pi}} \int_0^{\xi} e^{-\frac{y^2}{2}} dy \tag{4.34}$$

Momentum flux in terms of the mean velocity

The contributions of the fluctuating components of the velocity being small, the above expression for the mean flux of the momentum, Eq 4.34, can be rewritten as

$$\frac{u^2}{u_0^2} = \frac{1}{2}(1 + erf(\xi)) \tag{4.35}$$

The error function $erf(\xi)$ is defined as $\frac{2}{\sqrt{\pi}} \int_0^{\xi} e^{-y^2} dy$, and its values can be read from tables for specified arguments ξ.

4.4.2 Two-dimensional Jet

The governing equation of motion is identical to Eq 4.29 and the momentum transport relationship, Eq 4.30, is equally valid for a two-dimensional jet. However, the distribution of the momentum flux across the jet is given by $\overline{U^2} = \overline{U_m^2} f(\xi)$. Following a procedure akin to the case of the jet boundary layer, an ordinary differential equation for the universal function can be written as

$$f'' + \xi f' + f = 0 \tag{4.36}$$

Solution of the governing equation
On integrating Eq 4.36 twice, the function f is

$$f = C_1 e^{-\frac{\xi^2}{2}} + C_2 \tag{4.37}$$

$C_1 = 1$ and $C_2 = 0$ can be found using the boundary conditions; at $\xi = 0$, $f = 1$ and for $\xi \to \infty$, $f = 0$.

Therefore, for a two-dimensional turbulent jet in which the contribution of the fluctuating components is small, the distribution of the mean flux of the momentum is

$$\frac{u^2}{u_m^2} = e^{-\frac{\xi^2}{2}} \tag{4.38}$$

It is interesting to note that in the cases of the round jet and the two-dimensional slot, the universal distribution happens to be the same and is given by Eq 4.38. Further, for maintaining self-similarity of the turbulent flow, the characteristic width b must be a linear function of x. However, it is not clear from the integral analysis whether the constant of proportionality should be the same for both jets. This information has to be determined from experiments.

4.4.3 Characteristic Widths b and b_t

If the characteristic widths of spread of the momentum (or velocity differences) and the temperature differences are treated as equal, that is, $b = b_t$, the distributions of the momentum flux and the heat flux should turn out to be the same. In contrast, by Taylor's vorticity transfer theory, the intensity of the turbulent heat transfer is seen to be two times larger than the momentum transfer. Based on Prandtl's new theory, it has been established that (Abramovich, 1963)

$$\frac{\Delta T}{\Delta T_m} = \left(\frac{u}{u_m}\right)^{\frac{1}{\delta}} \tag{4.39}$$

where δ is the ratio of the transverse scales of the thermal and dynamic layers. The above result agrees with Taylor's formula when the exponent $\delta = 2$. This is true for the plane jet as well as the axially symmetric jet.

To comply with Taylor's theory, it will be interesting to find the ratio $\frac{b_t}{b}$ from the distributions given in Eq 4.23 and 4.14. Let us use Eq 4.39 to write the following,

$$\frac{\Delta T u}{\Delta T_m u_m} = \left(\frac{u}{u_m}\right)^{\frac{1}{\delta}} \frac{u}{u_m} = \left(\frac{u}{u_m}\right)^{\frac{\delta+1}{\delta}} = \left(\frac{u^2}{u_m^2}\right)^{\frac{\delta+1}{2\delta}} \tag{4.40}$$

Since $\frac{\Delta Tu}{\Delta T_m u_m} = \mathrm{e}^{-\left(\frac{\zeta^2}{2}\right)} = \mathrm{e}^{-\frac{1}{2}\left(\frac{r}{b_t}\right)^2}$, on selecting a value of $\delta = 2$, Eq 4.40 takes the form

$$\mathrm{e}^{-\left(\frac{r}{b_t}\right)^2} = \left(\frac{u^2}{u_m^2}\right)^{2\times\frac{3}{4}} = \mathrm{e}^{-\frac{3}{4}\left(\frac{r}{b}\right)^2} \tag{4.41}$$

Thus, $\left(\frac{b_t}{b}\right)^2 = \frac{4}{3}$ or $b_t = 1.15b$.

It is customary to write $b_t = \lambda b$. In the present analysis, we have shown that $\lambda = 1.15$ if Taylor's formula is considered valid. Though the λ values observed by different researchers vary over the range 1.12–1.2, the average comes out to be fairly close with the above predicted value. The agreement seems remarkable, notwithstanding the fact that the effects of the fluctuating components were ignored in the analysis.

4.4.4 Momentum Flux Distribution in the Flow-development Region

The expressions for the distribution of the momentum flux across the jet, developed so far, are valid in the fully developed region of the jet. The assumption of self-similarity is valid for $\frac{x}{d}$ exceeding a value of approximately 10. Based on chi-square and normal approximations to a noncentral chi-square distribution, Maxwell (1970) derived a distribution for the momentum flux in the developing flow region. He could obtain a closed form solution for the flux distribution downstream of the potential core. The concept of a point-source was adopted to synthesise the flux distribution for jets of finite size, relying on the method of superposition.

In case of an axially symmetric jet, the following relationship was found to hold good,

$$\frac{\overline{\rho u_m^2}}{\rho U_0^2} = 1 - \mathrm{e}^{-\left(\frac{d}{2Cx}\right)^2} \tag{4.42}$$

For a very large $\frac{x}{d}$, Eq 4.42 can be approximated by

$$\frac{\overline{\rho u_m^2}}{\rho U_0^2} = \left(\frac{d}{2Cx}\right)^2 \tag{4.43}$$

The above equation is equivalent to Eq 4.15, as $d^2 \propto A$ and Cx is denoted by b. However, there is a difference between the C values used in the said equations by a factor of $\sqrt{2}$, as Maxwell (1970) derived the distribution of the momentum flux in the form $\frac{u^2}{u_m^2} = \mathrm{e}^{-\left(\frac{y}{b}\right)^2} = \mathrm{e}^{-\left(\frac{y}{Cx}\right)^2}$.

By taking the logarithm of Eq 4.42, the following relationship can be written:

$$\frac{x}{d} = \frac{1}{2C}\left\{-ln\left(1 - \frac{\left(\overline{\rho u_m^2}\right)}{\left(\overline{\rho U_0^2}\right)}\right)\right\}^{-\frac{1}{2}} \tag{4.44}$$

A plot of $\frac{x}{d}$ against the bracketed quantity on the right-hand side of Eq 4.44 should be linear in the fully developed zone of the circular jet. Using the data of Alexander et al. (Maxwell, 1970), a value for $C = 0.0773$ was found. In addition, a virtual origin correction of $0.97d$ was deducted from the physically measured value of x. The plot was linear beyond a distance of about $4d$, thereby indicating that the potential core extends to a distance of $4d$. This is in variance with the general belief that the potential core extends up to a distance of $6.2d$ downstream of the nozzle. The spreading coefficient C and the virtual origin correction were then used to find the lateral distribution of the momentum flux in the region of flow establishment. In doing so, three types of approximations to the noncentral chi-square distribution were used. Beyond the potential core, these distributions gradually transform to describe the fully developed region of the jet. For $\frac{x}{d}$ exceeding 6, Maxwell's equation reduces to

$$\frac{\rho u^2}{\rho U_0^2} = 0.809\mathrm{e}^{-3.234\left(\frac{r^2}{d^2}\right)} \tag{4.45}$$

In the analysis propounded by Maxwell, the concept of point-source was, no doubt, introduced. However, he made use of a virtual origin correction and the chi-square distribution functions, which are difficult to adopt in practice. It will be demonstrated, in our subsequent discussions, how an equivalent square shape can replace a circular nozzle and simplify the process of superposition.

4.5 Summary

Based on Reichardt's momentum transport theory, the governing equation of motion for jets was transformed to a linear differential equation involving u^2. An analytical expression for the distribution of the momentum flux across a circular jet was derived. The same was extended to a point-source, based on which the distributions of the momentum-flux across plane and axially symmetric jets were reestablished using the superposition technique. The theoretical distribution of velocity found from the momentum transport theory is in good agreement with experimental data. The point-source method is simple compared to Tollmien's or Götler's methods.

The distribution of scalar quantities (the excess temperature or the concentration) was also derived based on Reichardt's hypothesis, albeit the length scale for the distribution of the scalar quantities is larger than that of the momentum flux distribution.

Example 4.1: Velocities were measured at different cross sections downstream of a circular nozzle. The half velocity width $b_{0.5}$ was found to vary linearly with the distance x and $b_{0.5} = 0.095x$. All the data points were seen to follow a Gaussian distribution. Find the following,

(i) the nondimensional distribution of the momentum flux and
(ii) the relationship between $b_{0.5}$ and the length scale b_R used in Reichardt's theory.

Solution:

(i). In terms of the half velocity, the Gaussian distribution is

$$\frac{u}{u_m} = \mathrm{e}^{-0.693\left(\frac{r}{b_{0.5}}\right)^2}$$

Hence, the distribution of the momentum flux is

$$\frac{\rho u^2}{\rho u_m^2} = \mathrm{e}^{-1.386\left(\frac{r}{b_{0.5}}\right)^2} \tag{I}$$

(ii) From Reichardt's momentum transport theory,

$$\frac{\rho u^2}{\rho u_m^2} = \mathrm{e}^{-\frac{1}{2}\left(\frac{r}{b_R}\right)^2} \tag{II}$$

Equating (I) with (II),

$$\frac{1.386}{b_{0.5}^2} = \frac{1}{2b_R^2}$$

or

$$b_R = 0.60 b_{0.5} = 0.60 \times 0.095x = 0.057x$$

Example 4.2: For a round turbulent jet, the mean momentum flux distribution is given in Eq 4.42 as

$$\frac{\rho u_m^2}{\rho U_0^2} = 1 - \mathrm{e}^{-\left(\frac{d}{2Cx}\right)^2}$$

Adopting the C value from Example 4.1, find the longitudinal distance at which $\frac{u_m^2}{U_0^2} = 0.98$.

Solution: $C = 0.057$. Hence, $1 - e^{-\left(\frac{1}{2\times 0.057(x/d)}\right)^2} = 0.98$

or

$$\left\{\frac{1}{0.114(x/d)}\right\}^2 = 3.912$$

That is, $\frac{1}{0.114(x/d)} = 1.977$

Hence,

$$\frac{x}{d} = \frac{1}{0.114 \times 1.977} = 4.43$$

Therefore, at $x \simeq 4.4d$, the velocity ratio $\frac{u_m}{U_0} = 0.99$. To avoid ambiguity, the length of the potential core could be defined as that value of x beyond which the maximum velocity on the jet axis, u_m, falls below $0.99U_0$.

Example 4.3: Two identical circular nozzles, 10 cm in diameter, discharge air at a velocity of 100 m/s. The centre-to-centre spacing of the nozzles is 30 cm and the two axes are parallel. Find the specific momentum flux and the u-component of the velocity at the point 'P', on the axis of symmetry, at a distance of 3 m downstream of the nozzle. Use $C = 0.057$ in the Reichardt's solution.

Solution: From Reichardt's theory, we have $u_m^2 = U_0^2\left\{\frac{A}{2\pi b^2}\right\}$

or $u_m^2 = U_0^2\left\{\frac{1}{8C^2\left(\frac{x}{d}\right)^2}\right\}$

Hence, at point 'A' on the axis of jet-1,

$$u_m^2 = U_0^2\left\{\frac{1}{8 \times 0.057^2\left(\frac{300}{10}\right)^2}\right\} = 0.04275 \times U_0^2$$

Further, from the relationship $u^2 = u_m^2 e^{-\frac{1}{2}\left(\frac{r}{b}\right)^2}$, at point '$P$', the value of u^2 contributed by jet-1 is

$$u^2 = u_m^2 e^{-\frac{1}{2}\left\{\frac{15}{0.057\times 300}\right\}^2} = 0.680u_m^2 = 0.680 \times 0.04275 \times U_0^2 = 0.02907 \times U_0^2$$

Similarly, the contribution of jet-2 at point 'P' will be $u^2 = 0.02907 \times U_0^2$. Hence, the resultant momentum flux due to the two jets will be the sum total of the two contributions.

Thus, the momentum flux at point 'P' $= \rho \times \{(0.02907 \times U_0^2) + (0.02907 \times U_0^2)\} = 0.05814 \times \rho \times U_0^2$.

Assuming a value for $\rho = 1.23\ \text{kg/m}^3$, the momentum flux at the point 'P' $= 0.05814 \times 1.23 \times 100^2 = 715.12$ Pa.

$$\text{The velocity } u = \sqrt{0.05814 \times 100^2} = 24.1\ \text{m/s}.$$

It is important to note that the squares of the velocities are added first to find the velocity at point 'P'. It is erroneous to add the two velocities to arrive at the resultant value.

5 Three-dimensional Free Jet: Single Outlet

5.1 Introduction

A jet emanating from a rectangular or any other noncircular shape of orifice is known as a three-dimensional jet. Therefore, flow from elliptical and triangular nozzle shapes, though not commonly used, will also be treated as three-dimensional free jets. Three-dimensional mixing can be observed in case of submerged outlets of cooling water systems in power plants, basin outlets of tidal power schemes, control gates of hydraulic structures, heating and ventilation of buildings, etc. As the three-dimensional flow field is complex, a complete theoretical solution is difficult to find. Most of the existing solutions are semi-empirical in nature. Therefore, the coefficients of the semi-empirical relationships need to be determined through experiments.

Investigations on three-dimensional free jets were initiated in the mid-sixties, notably by Yevdjevich (1966), Sforza et al. (1966), Trentacoste and Sforza (1967) and Sforza (1969). The investigations of Yevdjevich were primarily experimental in nature and involved slot jets of different aspect ratios, lying in the range of 4–94. Sforza derived theoretical velocity distributions based on Prandtl's eddy viscosity model and also conducted further experiments on three-dimensional jets. The experimental observations helped to demarcate three distinct zones in the flow field. Each displayed some specific features as regards the velocity decay and the growth of the shear layer.

Flow zones of a slender jet

To begin with, let us take the case of a slender jet, that is, a jet whose aspect ratio L_0/B_0 is say greater than 5. Schematic diagram of the resulting flow field is depicted in Fig 5.1. Here, u_{m0} represents the maximum velocity over a cross section of the jet and is a function of the distance x.

The flow field can be characterised by three regions indicated below:

(i) A potential core region (PC): This is where mixing initiated by the jet boundaries at the top and the bottom has not penetrated the entire thickness

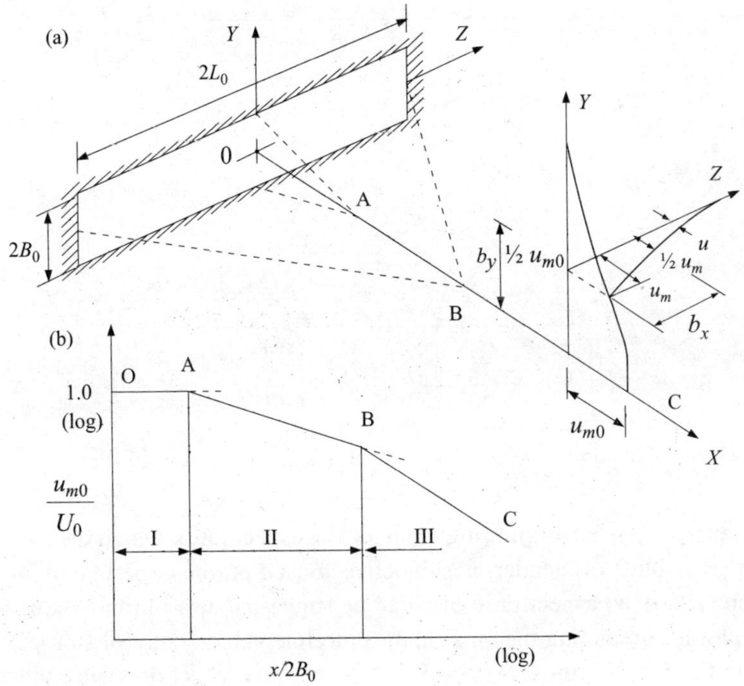

Figure 5.1 Schematic diagram of a three-dimensional slender jet

of the jet. A region close to the axis of the jet exists where the maximum velocity remains unaffected by the spreading of the shear layer. Hence, $\frac{u_{m0}}{U_0}$ ratio has a value of 1.0.

(ii) A characteristic decay region (CD): The zone lying between points 'A' and 'B' in Fig 5.1 represents the CD zone. The inner mixing boundaries from the two sides meet at point 'B'. The length of the CD region depends on the aspect ratio of the outlet. In this region, the velocity decay is akin to the two-dimensional jet, that is, $u_{m0} \sim x^{-\frac{1}{2}}$.

(iii) An axisymmetric decay region (AD): This is the farthest region where the maximum velocity u_{m0} decays linearly with the distance x, that is, $u_{m0} \sim x^{-1}$. However, the flow field in the AD region is not exactly symmetrical around the jet axis. Depending on the aspect ratio of the outlet, the flow behaves like a circular jet after traversing some distance downstream.

Bluff jets

Jets which have aspect ratios nearly equal to 1 are known as 'bluff jets'. On the other hand, 'slender jets' have aspect ratios which are significantly larger

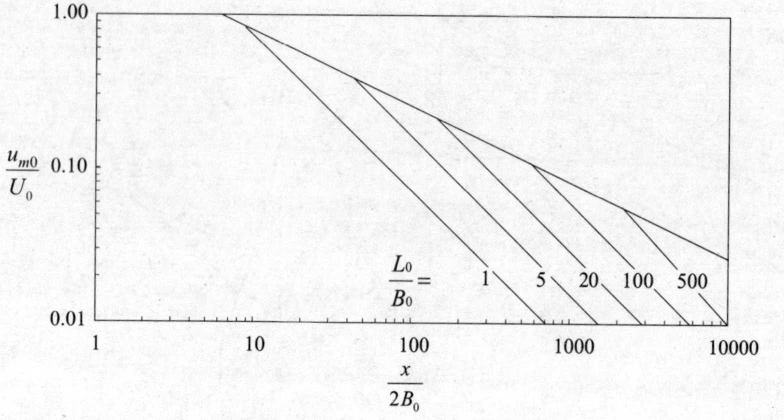

Figure 5.2 Decay of maximum velocity for slot jets (Source: Yevdjevich, 1966)

or smaller than 1. The exact limiting value of the aspect ratio, which decides whether a jet is bluff or slender, is subjective. Based on the experiments of Yevdjevich (1966), an aspect ratio of 5 can be suggested as a limiting value. A typical plot for predicting the maximum centreline velocity u_{m0} of slot jets is shown in Fig 5.2. The three regions PC, CD and AD are clearly visible.

In the case of the bluff jets, for example, jets issuing from a square nozzle or an equilateral triangle, the extent of the CD region is negligible. The axisymmetric decay region immediately follows the potential core region.

Effect of aspect ratio on CD region

Typically, in case of a jet having an aspect ratio of 5, the plane jet velocity decay behaviour extends from $\frac{x}{2B_0} = 6$–32. The upper limiting value of the CD region gets bigger for larger values of the aspect ratio.

The experimental results of Trentacoste and Sforza (1967) further revealed that the behaviour of slender jets did not exactly agree with the abovementioned scheme. For very slender jets having aspect ratios of 20 and 40, $u_{m0} \sim x^{-\frac{1}{2}}$ in the CD region. However, in the case of jets having an aspect ratio of 10, the exponent of x was different, and $u_{m0} \sim x^{-\frac{1}{3}}$. A similar feature was noticed even for an elliptic jet having an aspect ratio of 9.35.

5.2 Slender Jets: Velocity Distribution and Growth

Experiments show that the distribution of the mean velocity, u_m versus y, exhibits self-similarity for all x-locations lying beyond the potential core.

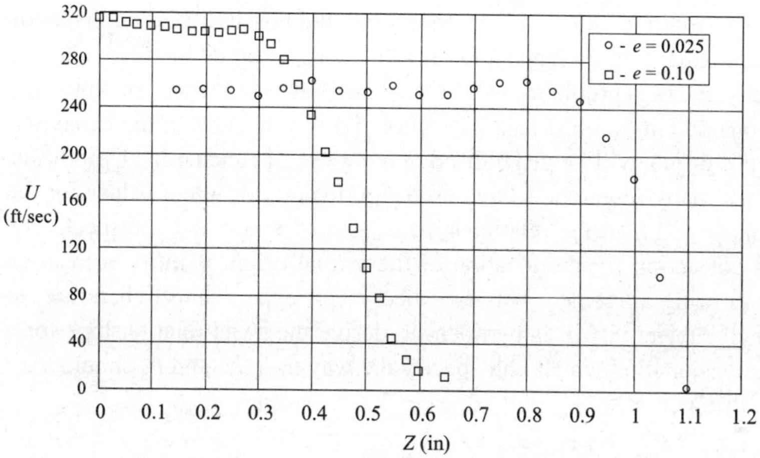

Figure 5.3 Saddle-shaped mean velocity distribution (Source: Trentacoste and Sforza, 1967)

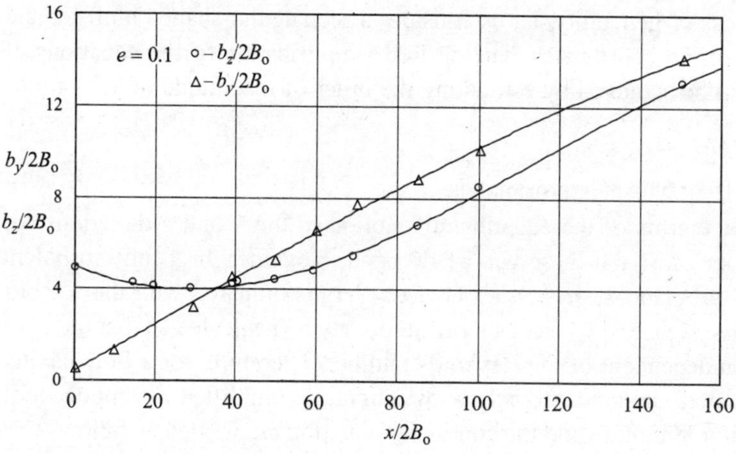

Figure 5.4 Growth of b_y and b_z for a slender jet (Source: Trentacoste and Sforza, 1967)

However, the distribution of u with z in the CD region displays a saddle shape. The maximum value of u occurs symmetrically on either side of the axis, a little away from the centreline, as shown in Fig 5.3.

In the AD region, the velocity distributions in both y and z directions have been observed to be self-similar. The length scales in the two directions b_y and b_z are generally different. The growth of the two scales with distance-x is shown in Fig 5.4. For very large values of x, the two length scales have a tendency to attain identical values. Moreover, the possibility of an inversion of the jet, that is, b_z growing at a faster rate than b_y, cannot be ruled out.

In many practical problems, we come across bluff jets. Hence, the following discussions pertain to bluff jets. In Section 5.5 of this chapter, the behaviour of all types of jets will be generalised making use of Reichardt's hypothesis. In the past, many researchers have used significantly different values for the exponents p and q in the relationships $u_{m0} \propto x^p$ and $b \propto x^q$, respectively. Often, a correction for the location of the virtual origin is introduced in an attempt to bring agreement between theory and experiments. Therefore, to avoid such subjectivity, it is pertinent to derive the basic relationships for a three-dimensional jet which will specify the way u_{m0}, b_y and b_z should vary with the distance, x.

5.3 Governing Laws for the Variation of Velocity and Length Scales

Without loss of generality, let us consider a rectangular-shaped bluff jet and adopt Cartesian coordinates. The desired simplification of the equations of motion can be achieved by estimating the order of magnitude of the various terms.

Boundary layer type of approximation

The various terms of the equations of motion in the y and z directions are of the order of δ, and these can be dropped altogether. In a fully turbulent flow, the turbulent shear stresses are very large compared with the viscous shear stress. Hence, the viscous terms, too, may be neglected so that the flow becomes independent of the Reynolds number. Therefore, for a flow having negligible pressure gradient in the x-direction, the simplified x-component of the equation of motion and the continuity equation are written as follows.

$$u\frac{\partial u}{\partial x} + v\frac{\partial u}{\partial y} + w\frac{\partial u}{\partial z} = -\left(\frac{\partial \overline{u'v'}}{\partial y} + \frac{\partial \overline{u'w'}}{\partial z}\right) \tag{5.1}$$

and

$$\frac{\partial u}{\partial x} + \frac{\partial v}{\partial y} + \frac{\partial w}{\partial z} = 0 \tag{5.2}$$

Equations 5.1 and 5.2 will be referred to as the boundary layer type of equations for a three-dimensional jet.

Self-similarity of mean velocity, *u*

Based on experiments on three-dimensional jets and the experience we have had with other types of jets, it looks logical to assume the velocity profiles in the y and z planes to be self-similar. The velocity scale and the length scale in the y direction are u_{m0} and b_y, respectively. The corresponding scales in the z direction are u_m and b_z, as shown in Fig 5.1. In the x–y plane passing through the axis of the jet, let us assume

$$\frac{u_m}{u_{m0}} = f_1\left(\frac{y}{b_y}\right) = f_1(\eta_y) \tag{5.3}$$

where u_m is the mean velocity in the x direction at a distance y from the axis of the jet. The velocity u_{m0} is the maximum value of u_m, and f_1 represents a functional relationship. Similarly, in the x–z planes, the velocity distribution can be described by

$$\frac{u}{u_m} = g_1\left(\frac{z}{b_z}\right) = g_1(\eta_z) \tag{5.4}$$

where u is the turbulent mean velocity in the direction x at a distance z from the axis, η_z is the nondimensional distance and g_1 represents a functional relationship.

Combining Eqs 5.3 and 5.4, the local mean velocity at any point can be expressed in terms of the maximum velocity on the axis of the jet. Thus, one can write the following relationship.

$$\frac{u}{u_{m0}} = f_1(\eta_y)g_1(\eta_z) \tag{5.5}$$

Self-similarity of *v* and *w*

Experience with plane and circular jets shows that v and w velocity components can be normalised by the u velocity scale (Anwar and Weller, 1969), and the complex operation of getting the dimensionless v and w values from the equation of continuity can be circumvented. Following such a procedure, let us assume

$$\frac{v}{u_m} = g_2(\eta_z) \tag{5.6}$$

and

$$\frac{w}{u_m} = g_3(\eta_z) \tag{5.7}$$

Combination of Eq 5.3 with Eqs 5.6 and 5.7, one at a time, results in

$$\frac{v}{u_{m0}} = f_1 g_2 \tag{5.8}$$

and

$$\frac{w}{u_{m0}} = f_1 g_3 \tag{5.9}$$

Self-similarity of turbulent shear stress

Once again, based on the observed properties of self-preserving jets (Wygnanski and Fiedler, 1969), one can write the shear stress components as

$$\frac{\overline{u'v'}}{u_m^2} = g_4(\eta_z) \tag{5.10}$$

and

$$\frac{\overline{u'w'}}{u_m^2} = g_5(\eta_z) \tag{5.11}$$

By combining Eqs 5.3, 5.10 and 5.11, we have the following.

$$\frac{\overline{u'v'}}{u_{m0}^2} = f_1^2 g_4 \tag{5.12}$$

$$\frac{\overline{u'w'}}{u_{m0}^2} = f_1^2 g_5 \tag{5.13}$$

Exponents p, q_1 and q_2

Let us suppose that the velocity and length scales can be expressed by the exponential relationships, $u_{m0} \propto x^p$, $b_y \propto x^{q_1}$ and $b_z \propto x^{q_2}$.

The purpose is to evaluate the unknown exponents p, q_1 and q_2. Let us use the expressions for the mean velocities and the shear stress components proposed in Eqs 5.12 and 5.13 in the boundary layer type equation. On substitution and by rearranging terms, from Eq 5.1 we can arrive at the following,

$$\begin{aligned} 2 f_1 f_1' g_4 = & -\frac{b_y}{b_z} f_1^2 g_5' - \frac{b_y u_{m0}'}{u_{m0}} f_1^2 g_1^2 + b_y' \eta_y f_1 f_1' g_1^2 \\ & + \frac{b_y b_z'}{b_z} \eta_z f_1^2 g_1 g_1' - f_1 f_1' g_1 g_2 - \frac{b_y}{b_z} f_1^2 g_1' g_3 \end{aligned} \tag{5.14}$$

where the primes on u_{m0}, b_y and b_z denote differentiation with respect to x, and the primes on f and g functions denote differentiation with respect to η_y and η_z, respectively.

From the property of similarity of velocity profiles, it is evident that the left-hand side of Eq 5.14 is independent of x. In order that the equation be valid at any section and is nontrivial, the right hand side of the equation must be independent of x. Hence, in the third term of the right hand side, we should have $b'_y \propto x^0$, that is, $q_1 = 1$. Similarly, in the last term on the right hand side, $\frac{b_y}{b_z} \propto x^0$. This results in $q_2 = q_1 = 1$. Other terms are unable to yield any additional information about the exponents already found.

To evaluate the other unknown exponent p, we need to take the help of the integral momentum equation

$$\frac{d}{dx}\int_{-\infty}^{\infty}\int_{-\infty}^{\infty} \rho u^2 dydz = 0 \tag{5.15}$$

Again, based on the principle of similarity, the integral momentum equation leads to

$$\frac{d}{dx}(\rho u_{m0}^2 b_y b_z) = 0 \tag{5.16}$$

Thus, $x^{2p+q_1+q_2} \propto x^0$, that is, $2p + q_1 + q_2 = 0$. Using the earlier results for q_1 and q_2, we have $p = -1$. Therefore, we can write the basic law for the variation of the velocity and the length scales as follows.

$$\begin{aligned} u_{m0} &\propto x^{-1} \\ b_y &\propto x \\ b_z &\propto x \end{aligned} \tag{5.17}$$

5.3.1 Dimensional Analysis

From the observations on round jets, it has been well established that the initial momentum flux M_0 is a dominant parameter which governs the behaviour of jets. Therefore, for three-dimensional jets, we shall take the liberty to express the functional relationships as

$$u_{m0} = \phi_1(M_0, \rho, x) \tag{5.18}$$

and

$$b_y, b_z = \phi_2(M_0, \rho, x) \tag{5.19}$$

From Eq 5.18, we can get an expression for the decay of the maximum streamwise velocity as

$$\frac{u_{m0}}{U_0} = \frac{C_u}{(\frac{x}{(\sqrt{A})}} \tag{5.20}$$

Similarly, the length scales work out as

$$\frac{b_y}{x} = C_1$$
$$\frac{b_z}{x} = C_2 \tag{5.21}$$

In the above expressions, the distance x is to be measured from the respective virtual origins of u_{m0}, b_y and b_z.

5.4 Application of Momentum Transport Hypothesis

From experiments, Sforza (1969) observed that in the far field, the mean velocity components in the y–z planes can be described by ellipses. These were transformed to circular shapes by taking the ratio of the radius vector r of any isovel to the radius vector r_0 of a reference isovel. The radius vector r_0 was considered as a function of both x and the orientation θ. The equations of motion were solved by assuming a modified eddy viscosity relationship in terms of r_0. Generally speaking, the form of the velocity distribution was quite similar to the one meant for axially symmetric jets. Nonetheless, his analysis could incorporate the three-dimensional jet data in determining the constants of integration.

The mean velocity field of a three-dimensional free jet can be predicted in a simple way by using the point-source, as defined in Eq 4.16. The application of Reichardt's hypothesis to the case of a three-dimensional jet can be done in two distinct ways. Both methods are based on the principle of superposition. The first method is quite similar to the one used for a round jet as described in Section 4.2, Eq 4.15. The second method relies on the point-source concept, entrapped in Eq 4.16, and will have certain inherent advantages.

5.4.1 Twin Mixing Length Approach

The velocity profiles in the x–y and the x–z planes of three-dimensional jets have been observed to be self-similar. This is true if appropriate length scales b_y and b_z are selected to form the nondimensional distances $\eta_y = \frac{y}{b_y}$ and $\eta_z = \frac{z}{b_z}$. In general, the rates at which the jet grows in the y and z directions are quite different. This is so, even in the AD region. Only in the very far field, the rate of growth tends to be the same in all the directions.

Since our interest often lies in the near field of the AD region, it seems reasonable to select two different mixing lengths which are both dependent on x alone (Pani and Das, 1983).

Governing equations

The simplified governing equations of motion for a steady three-dimensional turbulent jet in a zero pressure gradient field have been framed following a boundary layer type of approximation. In terms of the instantaneous velocities, the momentum equation is

$$U\frac{\partial U}{\partial x}+V\frac{\partial U}{\partial y}+W\frac{\partial U}{\partial z}=0 \qquad (5.22)$$

and the continuity equation is given by

$$\frac{\partial U}{\partial x}+\frac{\partial V}{\partial y}+\frac{\partial W}{\partial z}=0 \qquad (5.23)$$

Combination of Eqs 5.22 and 5.23, followed by time averaging of various terms, leads to

$$\frac{\partial \overline{U^2}}{\partial x}+\frac{\partial \overline{UV}}{\partial y}+\frac{\partial \overline{UW}}{\partial z}=0 \qquad (5.24)$$

Reichardt's hypothesis in *y* and *z* directions

As the jet grows at different rates in the y and z directions, two mixing length parameters Λ_1 and Λ_2 are selected to take care of the momentum transfer in the two mutually perpendicular directions. Hence, extending Reichardt's hypothesis in two mutually perpendicular directions y and z, we can write

$$\overline{UV}=-\Lambda_1\frac{\partial \overline{U^2}}{\partial y} \qquad (5.25)$$

and

$$\overline{UW}=-\Lambda_2\frac{\partial \overline{U^2}}{\partial z} \qquad (5.26)$$

where Λ_1 and Λ_2 are linear functions of x. Using Eqs 5.25 and 5.26 in Eq 5.24, we have

$$\frac{\partial \overline{U^2}}{\partial x}=\Lambda_1\frac{\partial^2 \overline{U^2}}{\partial y^2}+\Lambda_2\frac{\partial^2 \overline{U^2}}{\partial z^2} \qquad (5.27)$$

Self-similarity of momentum flux

In the axisymmetric decay region, the momentum flux profiles in the y and z directions have been observed to satisfy self-similarity. Therefore, let us assume the distributions as

$$\frac{\overline{U_m^2}}{\overline{U_{m0}^2}}=f(\xi) \qquad (5.28)$$

and

$$\frac{\overline{U2}}{\overline{U_m^2}} = g(\zeta) \tag{5.29}$$

in which the dimensionless coordinates $\xi = \frac{y}{B_y}$ and $\zeta = \frac{z}{B_z}$; B_y and B_z are the characteristic scales for the distribution of the instantaneous momentum flux. By combining the above two equations, we have

$$\frac{\overline{U^2}}{\overline{U_{m0}^2}} = f(\xi)g(\zeta) \tag{5.30}$$

Normalised momentum transport equation

Using Eq 5.30 in Eq 5.27, an expression can be derived in the following form

$$fg - N_1\xi f'g - N_2\zeta fg' = N_3 f''g + N_4 fg'' \tag{5.31}$$

where

$$N_1 = \frac{\overline{U_{m0}^2}}{\frac{\partial \overline{U_{m0}^2}}{\partial x}} \frac{B_y'}{B_y} \tag{5.32}$$

$$N_2 = \frac{\overline{U_{m0}^2}}{\frac{\partial \overline{U_{m0}^2}}{\partial x}} \frac{B_z'}{B_z} \tag{5.33}$$

$$N_3 = \frac{\Lambda_1 \overline{U_{m0}^2}}{\frac{\partial \overline{U_{m0}^2}}{\partial x}} \frac{1}{B_y^2} \tag{5.34}$$

$$N_4 = \frac{\Lambda_2 \overline{U_{m0}^2}}{\frac{\partial \overline{U_{m0}^2}}{\partial x}} \frac{1}{B_z^2} \tag{5.35}$$

For a self-preserving flow, N_1, N_2, N_3 and N_4 must be constants. The same should also be true for the ratios $\frac{N_1}{N_3}$ and $\frac{N_2}{N_4}$. Therefore, $\Lambda_1 = B_y B_y'$ and $\Lambda_2 = B_z B_z'$.

Further, we notice that Λ_1 and Λ_2 are linear functions of x since both B_y and B_z vary linearly with the distance x. Substituting for Λ_1 and Λ_2 in Eq 5.31, we can arrive at the following:

$$g\frac{B_y'}{B_y}(f + \xi f' + f'') + f\frac{B_z'}{B_z}(g + \zeta g' + g'') = 0 \tag{5.36}$$

5.4.2 Specific momentum-flux distribution

In order to satisfy Eq 5.36, each of the two bracketed terms on the left hand side can be equated to zero and solved independently. The boundary conditions to be satisfied by the f and g functions are

$$\begin{aligned} \xi &= 0; \quad f(\xi) = 1 \\ \xi &= \infty; \quad f(\xi) = 0 \\ \zeta &= 0; \quad g(\zeta) = 1 \\ \zeta &= \infty; \quad g(\zeta) = 0 \end{aligned} \tag{5.37}$$

The solutions for the f and g functions are

$$\begin{aligned} f &= \frac{\overline{U_m^2}}{\overline{U_{m0}^2}} = \mathrm{e}^{-\frac{\xi^2}{2}} \\ g &= \frac{\overline{U^2}}{\overline{U_m^2}} = \mathrm{e}^{-\frac{\zeta^2}{2}} \\ f.g &= \frac{\overline{U^2}}{\overline{U_{m0}^2}} = \mathrm{e}^{-\left(\frac{\xi^2+\zeta^2}{2}\right)} \end{aligned} \tag{5.38}$$

5.4.3 Decay of maximum momentum flux

The way the maximum value of the momentum flux drops along the x-axis can be worked out from the momentum conservation principle. For a negligible pressure gradient case, the momentum is conserved and can be expressed as

$$\int_{-\infty}^{\infty}\int_{-\infty}^{\infty} \rho\overline{U^2}dydz = \rho\overline{U_0^2}A \tag{5.39}$$

From Eqs 5.38 and 5.39, the following expression shows the decay of maximum instantaneous momentum flux,

$$\frac{\overline{U_{m0}^2}}{\overline{U_0^2}} = \frac{A}{2\pi B_y B_z} \tag{5.40}$$

Momentum flux based on mean velocity

The mean as well as the fluctuating motions contribute to the instantaneous momentum flux, $\rho\overline{U^2}$. The contribution of the fluctuating velocity component is an order of magnitude smaller than that of the mean components.

Therefore, let us assume $\overline{U^2} = u^2$. However, this assumption might prove erroneous near the edge of the jet.

For all practical purposes, from Eq 5.40, the decay of the mean momentum flux can be expressed as

$$\frac{u_{m0}^2}{U_0^2} = \frac{A}{2\pi B_y B_z} \tag{5.41}$$

5.4.4 Decay of maximum mean velocity

The variation of u_{m0} with distance x can be obtained from Eq 5.41 as

$$\frac{u_{m0}}{U_0} = \left(\frac{A}{2\pi B_y B_z}\right)^{\frac{1}{2}} \tag{5.42}$$

Further, the velocity distributions in the x–y and x–z planes are expressed as

$$\begin{aligned} \frac{u_m}{u_{m0}} &= \mathrm{e}^{-\frac{\xi^2}{4}} \\ \frac{u}{u_m} &= \mathrm{e}^{-\frac{\zeta^2}{4}} \\ \frac{u}{u_{m0}} &= \mathrm{e}^{-\frac{(\xi^2+\zeta^2)}{4}} \end{aligned} \tag{5.43}$$

Velocity distribution based on half velocity length scale

It will be convenient to use the half widths b_y and b_z in the integral analysis, which can help to adopt the experimental coefficients reported by many other researchers. The normal distribution described in Eq 5.43 will be compatible with the distribution of momentum flux if $b_y = 1.665B_y$ and $b_z = 1.665B_z$. One can also rewrite Eq 5.43 in the traditional form making use of the half widths such as

$$\begin{aligned} \frac{u_m}{u_{m0}} &= \mathrm{e}^{-0.693\eta_y^2} \\ \frac{u}{u_m} &= \mathrm{c}^{-0.693\eta_z^2} \\ \frac{u}{u_{m0}} &= \mathrm{e}^{-0.693(\eta_y^2+\eta_z^2)} \end{aligned} \tag{5.44}$$

where $\eta_y = \frac{y}{b_y}$ and $\eta_z = \frac{z}{b_z}$.

On substitution for B_y and B_z in terms of b_y and b_z respectively, the maximum velocity decay relationship from Eq 5.42 works out to be

$$\frac{u_{m0}}{U_0} = 0.664\left(\frac{A}{b_y b_z}\right)^{\frac{1}{2}} \tag{5.45}$$

In the axisymmetric decay region of three-dimensional jets, the length scales vary linearly with the distance x. Let us, therefore, set

$$\begin{aligned} b_y &= c_1 x \\ b_z &= c_2 x \end{aligned} \tag{5.46}$$

By combining Eqs 5.45 and 5.46, the decay of the maximum mean velocity can be expressed as

$$\frac{u_{m0}}{U_0} = \frac{0.664}{c}\left(\frac{x}{\sqrt{A}}\right)^{-1} \tag{5.47}$$

where $c = \sqrt{c_1 c_2}$.

$\sqrt{A}$, the new characteristic length

It may be noticed that $\sqrt{A}$ appears as a characteristic length parameter in Eq 5.47 and is also in agreement with the results of the dimensional analysis. *A limitation of Eq 5.47 is that it can only be applied in the AD region where the similarity of the momentum flux profiles was assumed to be valid.* Nonetheless, Eq 5.38 does offer a big advantage as it allows superposition of the momentum flux from multiple outlets. Details of this aspect will be presented in Chapter 6.

5.5 Unified Solution Based on Point-Source

In a jet, the potential core region and the characteristic decay region together are designated as the developing zone. To determine the distribution of momentum flux in the developing zone, the above method is not suitable. Even empirical fits lack generality as the coefficients need to be varied for bringing agreement between different sets of observations.

Single mixing length approach

This section describes the way a single mixing length can very effectively be used for the derivation of the momentum flux distribution, in the most general way. *There will be no need to treat the flow domains like the potential core, the characteristic decay and the axisymmetric decay regions as different entities.*

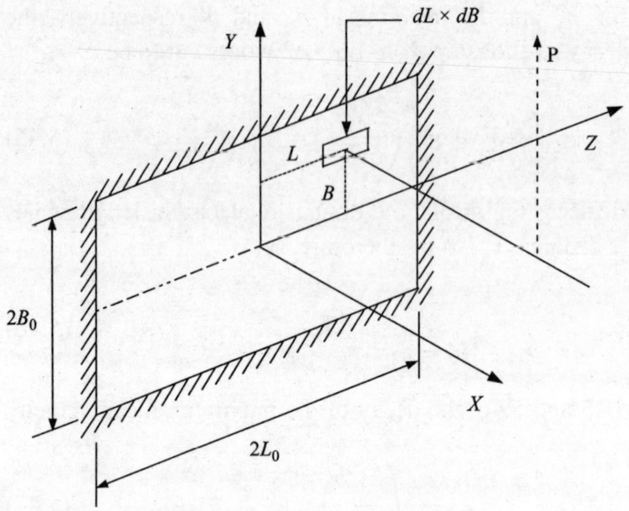

Figure 5.5 Definition sketch of a point-source

In addition, the difficulty faced in locating different types of virtual origins, associated with the velocity decay and the growth rates, can be eliminated.

Consider a point 'P', Fig 5.5, in the Y–Z plane at a distance x from the plane of the rectangular outlet of size $2B_0 \times 2L_0$. Let the coordinates of 'P' be y and z, measured from the axis of symmetry. Further, the elemental source dA is located at distances of L and B from the x-axis of the jet in the z and y directions, respectively. The momentum flux at point 'P', due to the existence of momentum flux through the elementary area dA, can be obtained from Eqs 4.14 and 4.16 as

$$\frac{u^2}{U_0^2} = \frac{e^{-\frac{1}{2}\left(\frac{r}{b}\right)^2}}{2\pi b^2} dL dB \tag{5.48}$$

Substituting for $r^2 = (z - L)^2 + (y - B)^2$ in Eq 5.48, the flux of the momentum generated at the point 'P' due to the total flow through the opening $2L_0 \times 2B_0$ is given by

$$\frac{u^2}{U_0^2} = \int_{-L_0}^{L0} \int_{-B_0}^{B0} \frac{e^{-\frac{1}{2}\left\{\left(\frac{z-L}{b}\right)^2 + \left(\frac{y-b}{b}\right)^2\right\}}}{2\pi b^2} dL dB \tag{5.49}$$

Closed form solution for velocity decay

Let us substitute

$$\frac{(y-B)}{\sqrt{2}b} = Y$$
$$\frac{(z-L)}{\sqrt{2}b} = Z \tag{5.50}$$

in Eq 5.49 to get

$$\frac{u^2}{U_0^2} = \frac{1}{4}\int \frac{2}{\sqrt{\pi}}e^{-Y^2}dY \int \frac{2}{\sqrt{\pi}}e^{-Z^2}dZ \tag{5.51}$$

The limits of integration for Y and Z are from $\left(\frac{y+B_0}{\sqrt{2}b}\right)$ to $\left(\frac{y-B_0}{\sqrt{2}b}\right)$ and from $\left(\frac{z+L_0}{\sqrt{2}b}\right)$ to $\left(\frac{z-L_0}{\sqrt{2}b}\right)$, respectively. Therefore, on integration, Eq 5.51 will yield the expression

$$\frac{u^2}{U_0^2} = \frac{1}{4}\left\{erf\left(\frac{y-B_0}{\sqrt{2}b}\right) - erf\left(\frac{y+B_0}{\sqrt{2}b}\right)\right\}$$
$$\times\left\{erf\left(\frac{z-L_0}{\sqrt{2}b}\right) - erf\left(\frac{z+L_0}{\sqrt{2}b}\right)\right\} \tag{5.52}$$

Along the longitudinal axis of the jet, $y = 0$ and $z = 0$. Hence, the maximum value of the momentum flux which occurs on the axis is given by

$$\frac{u_{m0}^2}{U_0^2} = erf\left(\frac{B_0}{\sqrt{2}b}\right) erf\left(\frac{L_0}{\sqrt{2}b}\right) \tag{5.53}$$

The decay of the maximum velocity along the x-axis is obtained by taking the square root of the above expression, and it results in the following,

$$\frac{u_{m0}}{U_0} = \left\{erf\left(\frac{B_0}{\sqrt{2}b}\right) erf\left(\frac{L_0}{\sqrt{2}b}\right)\right\}^{\frac{1}{2}} \tag{5.54}$$

If one were to find the distribution of the velocity u in terms of the maximum velocity u_{m0}, it can easily be done by taking the ratio of the square roots of $\frac{u^2}{U_0^2}$, given in Eq 5.52, and $\frac{u_{m0}^2}{U_0^2}$ of Eq 5.53.

5.6 Temperature Distribution

The distribution of the heat flux was derived in Section 4.3, Eq 4.23. Ignoring smaller order terms, the left-hand side of Eq 4.25 can be rewritten in terms of the mean velocity and mean temperature excess as

$$\frac{u_m \overline{\Delta T_m}}{U_0 \overline{\Delta T_0}} = \frac{A}{2\pi b_t^2} \tag{5.55}$$

Consider an elemental area dA in lieu of 'A' in the above expression. By combining Eq 5.55 with the heat flux distribution as given in Eq 4.23, one can write

$$\frac{u\overline{\Delta T}}{U_0 \overline{\Delta T_0}} = \frac{\mathrm{e}^{-\frac{\zeta^2}{2}} dA}{2\pi b_t^2} = \frac{\mathrm{e}^{-\frac{\zeta^2}{2}} dL dB}{2\pi b_t^2} \tag{5.56}$$

The heat flux at a point 'P' of Fig 5.5 is obtained by summing up the contributions from various elemental areas spread over the three-dimensional outlet. The steps involved are similar to the case of the momentum flux distribution. Therefore, on integration, the variation of the heat flux over the cross section of the jet is described by the following expression.

$$\frac{u\overline{\Delta T}}{U_0 \overline{\Delta T_0}} = \frac{1}{4}\left\{erf\left(\frac{y - B_0}{\sqrt{2}b_t}\right) - erf\left(\frac{y + B_0}{\sqrt{2}b_t}\right)\right\} \times \left\{erf\left(\frac{z - L_0}{\sqrt{2}b_t}\right) - erf\left(\frac{z + B_0}{\sqrt{2}b_t}\right)\right\} \tag{5.57}$$

If required, the temperature distribution $\frac{\overline{\Delta T}}{\overline{\Delta T_0}}$ can be found from Eq 5.57 by inserting the value for the ratio $\frac{u}{U_0}$ from Eq 5.52. The usage of the expressions for finding the distribution of the temperature is quite straightforward. Nonetheless, an error function calculator or table will be needed.

5.7 Experimental Observations

In the past, different aspects of three-dimensional jets have been observed based on experiments. These are summarised in the following sections.

5.7.1 Self-similarity

The distribution of the mean velocity in the x–y plane, measured at various downstream locations of a three-dimensional free jet, is shown in Fig 5.6. The outlet was rectangular and had an aspect ratio equal to 0.66. It can be observed that for $\frac{x}{\sqrt{A}}$ equal to or exceeding 10, the profiles satisfy self similarity.

The theoretical distribution given in Eq 5.44 is seen to be in close agreement with the observed values. In Fig 5.6, the theoretical relationship

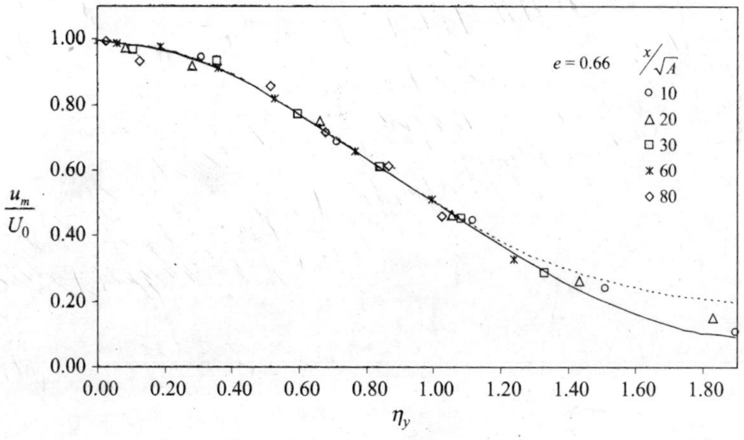

Figure 5.6 Mean velocity distribution for a three-dimensional jet (Source: Pani and Dash, 1983)

of Sforza (1969) is also shown as a dotted line. The two distributions are almost identical for $\eta_y \leq 1.0$. For larger values of η_y, they deviate from the experimental values by more or less the same margin. Furthermore, the experimental results of Sforza, not shown here, showed good agreement with the theoretical distribution derived from Reichardt's hypothesis. The velocity distribution in the x–z plane also depicts similar agreement between theory and experiments.

5.7.2 Spread Coefficients

The values of the coefficients c_1, c_2 and c in Eq 5.47 were found from the velocity profiles measured at various x stations. Rectangular nozzles of different aspect ratios were used in the experiment. As an illustration, the results for an aspect ratio of 10 are shown in Fig 5.7. In the characteristic decay region, which extends to a large distance x, the coefficients c_1 and c_2 change rapidly. Nonetheless, they tend to attain a constant value in the far field zone.

Despite the rapid changes in c_1 and c_2, the coefficient c $(= \sqrt{c_1 c_2})$ attains a near-constant value from the very beginning of the AD zone. The start of the AD region, however, depends upon the aspect ratio of the outlet. For larger values of the aspect ratio, the AD zone begins farther away from the nozzle. It was observed that c_1 and c_2 are nearly the same for a 'bluff' jet, as shown in Fig 5.7, and the CD region tends to shrink (Pani and Dash, 1983). For other values of aspect ratios, such as 40, 10, 1 and 0.66, the c values in the AD region were observed to be 0.10, 0.09, 0.11 and 0.104, respectively. As no systematic

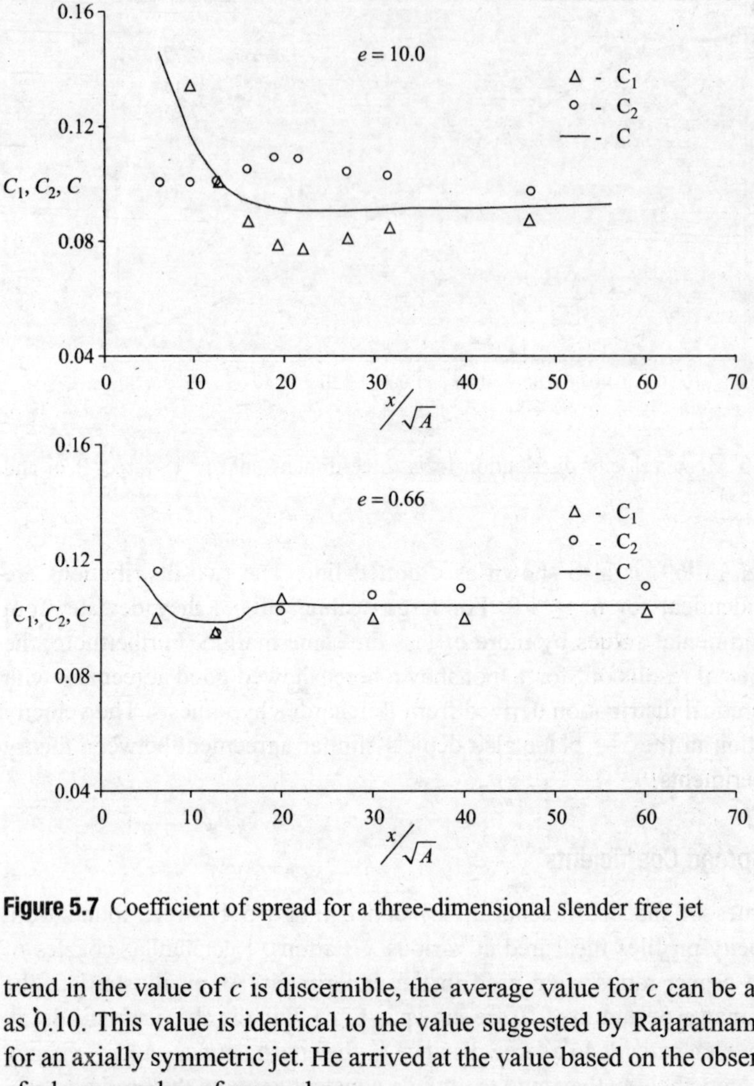

Figure 5.7 Coefficient of spread for a three-dimensional slender free jet

trend in the value of c is discernible, the average value for c can be assumed as 0.10. This value is identical to the value suggested by Rajaratnam (1976) for an axially symmetric jet. He arrived at the value based on the observations of a large number of researchers.

5.7.3 Decay of Maximum Velocity

By assuming a value $c = 0.10$ in Eq 5.47, a general expression for the decay of the maximum velocity along the axis of the jet can be written as

$$\frac{u_{m0}}{U_0} = 6.64\left(\frac{x}{\sqrt{A}}\right)^{-1.0} \tag{5.58}$$

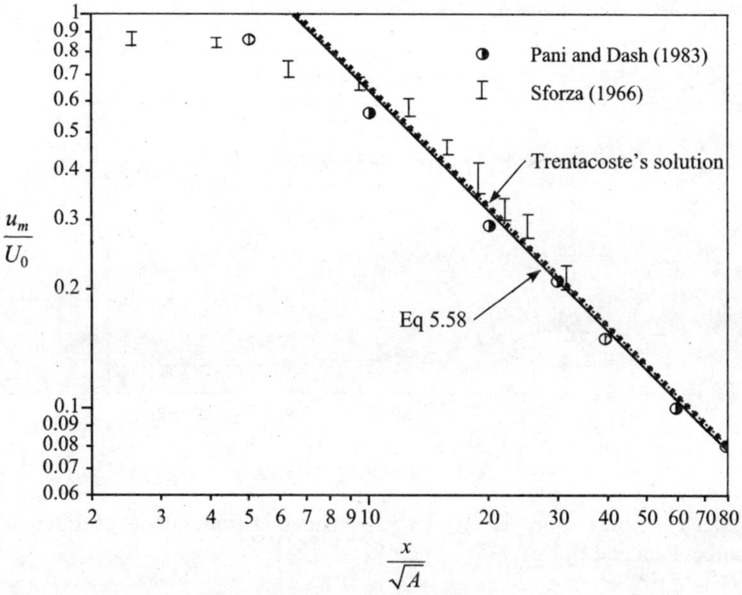

Figure 5.8 Decay of maximum velocity with distance normalised by $\sqrt{A}$ (Source: Pani and Dash, 1983)

For three-dimensional jets, Eq 5.58 was also proposed by Pani (1972) as an empirical relationship. Figure 5.8 shows the plot of Eq 5.58, together with the theoretical solution of Trentacoste (1967).

The bandwidth of scatter of the data points for an aspect ratio of 10 is shown in Fig 5.8. Interestingly, the data of Pani and Dash (1983) for a bluff jet seems to agree reasonably well with the results of the slender jet. The small difference between Eq 5.58 and Sforza's relationship can vanish if, instead of taking $c = 0.1$, we adopt a value of $c = 0.093$ as assumed by him.

Nozzle shape effect

The available experimental results for different shapes of orifices are shown in Fig 5.9 in the form $\frac{u_{m0}}{U_0}$ versus $\frac{x}{\sqrt{A}}$. The data includes rectangular outlets that had a wide range of values for the aspect ratio. It is evident that Eq 5.58 describes the observed values quite well, especially in the axisymmetric decay region.

5.7.4 Additional Observations on Axially Symmetric Jets

The results discussed above are applicable in the AD region. As an alternative, a single reliable value of c pertaining to an axially symmetric jet will be needed

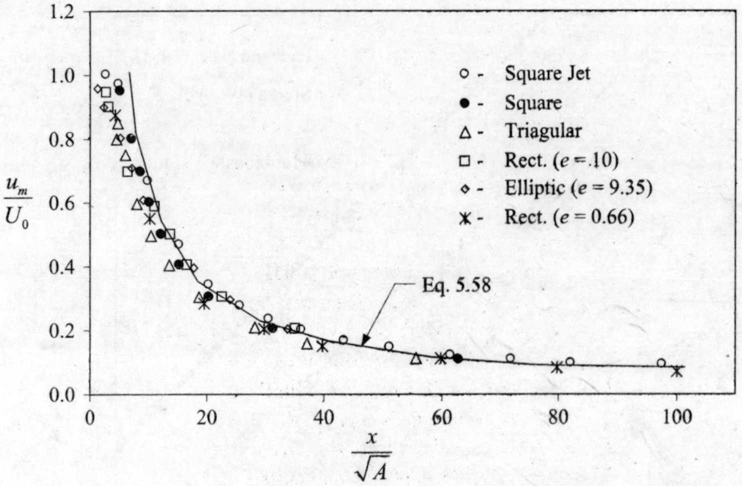

Figure 5.9 Decay of maximum velocity for jets emanating from outlets of different shapes (Source: Pani and Dash, 1983)

to apply Reichardt's theory for all the three flow regions, beginning from the outlet. Moreover, for superposition, it will be mathematically easier to work with simpler geometries like a square or a rectangle, in lieu of a circular shape. Experiments were conducted to provide the requisite inputs to Reichardt's theoretical relationships (Parameswaran, 1986). The variation of the maximum velocity with the streamwise distance $\frac{x}{d}$ is shown in Fig 5.10, making use of data from different sources.

Maxwell's relationship, Eq 4.42, and the conventional maximum velocity decay relationship, Eq 5.58, merge for $\frac{x}{d}$ exceeding 15. Both relationships agree very well with the observed values of Hinze (1959) and Parameswaran (1986).

Equivalence of circular and square outlets

While deriving the expression for the momentum flux of a point-source, the elemental area dA was replaced by an equivalent element of size $dL \times dB$, since the geometry of the source is unimportant. To validate this postulate, the circular shape was replaced with an equivalent square nozzle, and the theoretical velocity decay curve was plotted based on Eq 5.54. The observed data of the axially symmetric jet are replotted as in Fig 5.11. A correction 'a' for the location of the virtual origin is included in the plot to compare the experimental results from different sources.

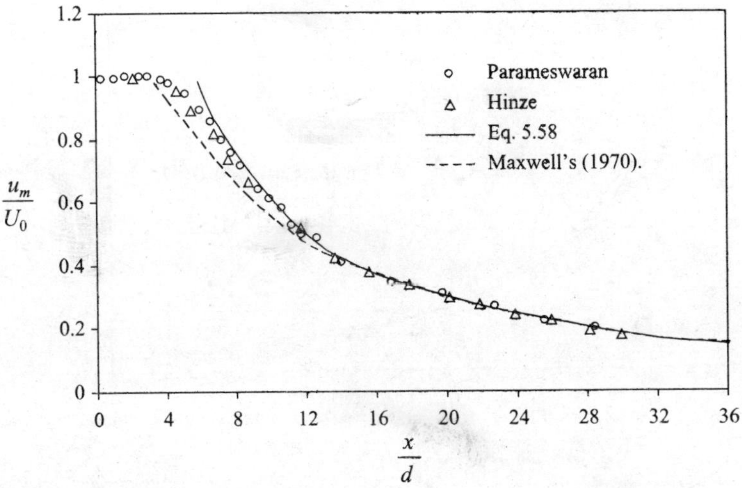

Figure 5.10 Decay of maximum velocity for a circular jet (Source: Parameswaran, 1986)

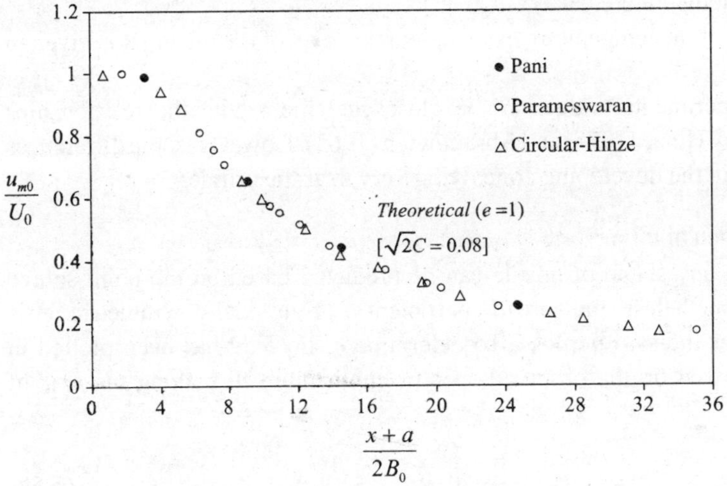

Figure 5.11 Equivalence of circular and square outlets

Good agreement is seen to exist between the theoretical curve for a square nozzle and the observed values of the circular jet. Therefore, *to derive theoretical velocity decay relationships it is enough to work with simple rectangular geometries. Such a step does not effect the accuracy of the results.*

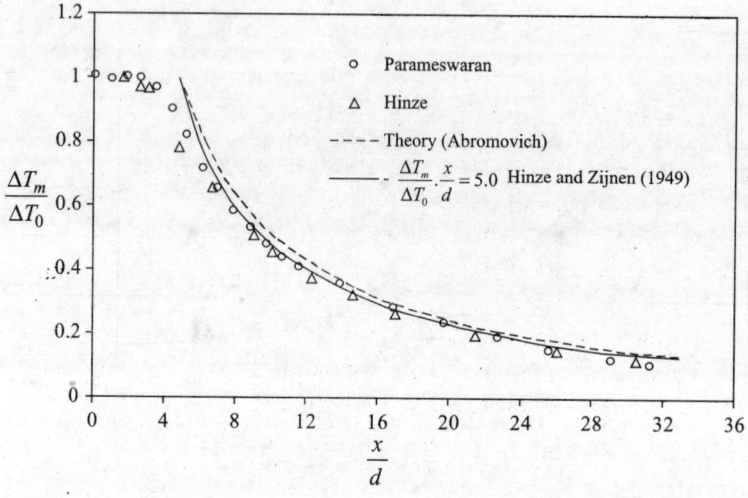

Figure 5.12 Axial variation of the excess temperature for a circular jet

Maximum temperature decay

The decay of the temperature excess along the axis of the round jet is given in Fig 5.12.

The experimental values are in close agreement with the relationships proposed by Hinze (1975) and Abramovich (1963). However, some differences do persist in the developing zone, lying very near the orifice.

Determination of c

Flow from any shape of nozzle can be predicted based on the point-source results. Nonetheless, the spread coefficient c of an axially symmetric jet is required for the said purpose. To determine c, Eq 5.59 has been plotted in Fig 5.13 based on the observed specific momentum flux along the axis of the jet.

$$\frac{x}{d} = \frac{1}{2\sqrt{2}c}\left\{-ln(1 - \frac{u_m^2}{U_0^2}\right\}^{-\frac{1}{2}} \tag{5.59}$$

From the slope of the linear relationship given in Eq 5.59, the value of $\sqrt{2}c$ equals 0.08. Following a similar approach, the temperature excess for the case of heated jets can be rewritten as

$$\frac{x}{d} = \frac{1}{2\sqrt{2}c\lambda}\left\{-ln(1 - \frac{\Delta T_m^2}{\Delta T_0^2)}\right\}^{-\frac{1}{2}} \tag{5.60}$$

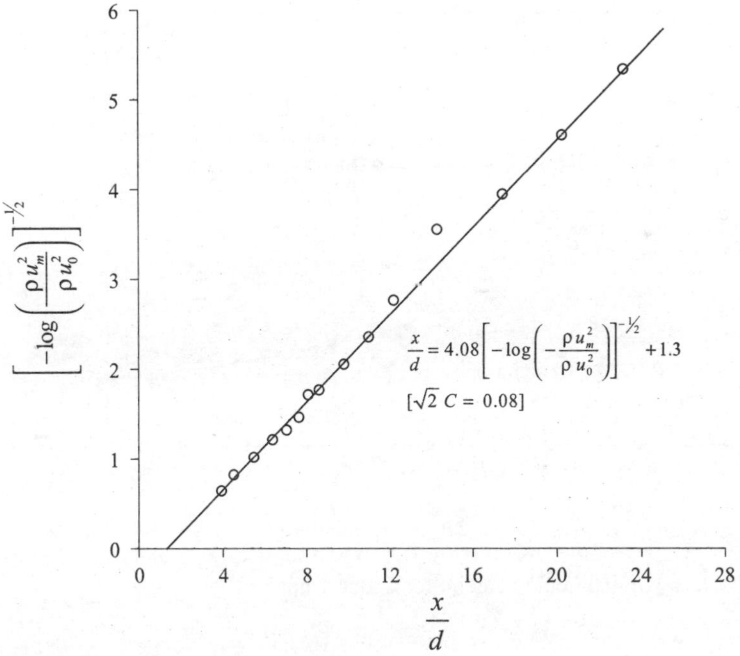

Figure 5.13 Determination of c for a circular jet

A plot of the bracketed quantity for various values of $\frac{x}{d}$ yields the slope of the linear relationship $\sqrt{2}c\lambda = 0.094$. As $\sqrt{2}c = 0.08$, the spread coefficient $\lambda = 1.17$. A similar analysis with the observed values of the momentum flux and the temperature excess of the square jet resulted in $\sqrt{2}c = 0.085$ and $\lambda = 1.136$. These values are in agreement with the results of many other researchers.

To generalise the point-source approach, which can be applied to all types of jets, including the three-dimensional geometry, a single value of $\sqrt{2}c$ should be used. This has been followed in the theoretical computations presented herein. The decay of the maximum velocity u_m along the axis of a square jet, shown in Fig 5.14, correspond to $\sqrt{2}c$ equals 0.08. The experimental results of Parameswaran (1986), Pani (1972) and Sforza et al. (1966) agree quite well with the values predicted from Eq 5.54.

Following a similar approach, the observed decay of the temperature excess along the axis of a square jet, as shown in Fig 5.15, is in agreement with the theoretical relation described by Eq 5.57.

The most interesting aspect of the point-source technique is that *no separate equation is necessary for predicting the maximum velocity in the developing*

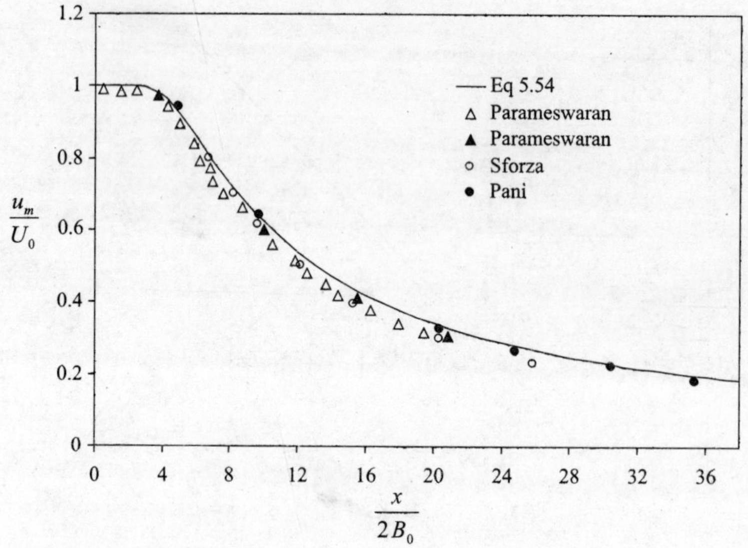

Figure 5.14 Decay of maximum velocity for a square nozzle

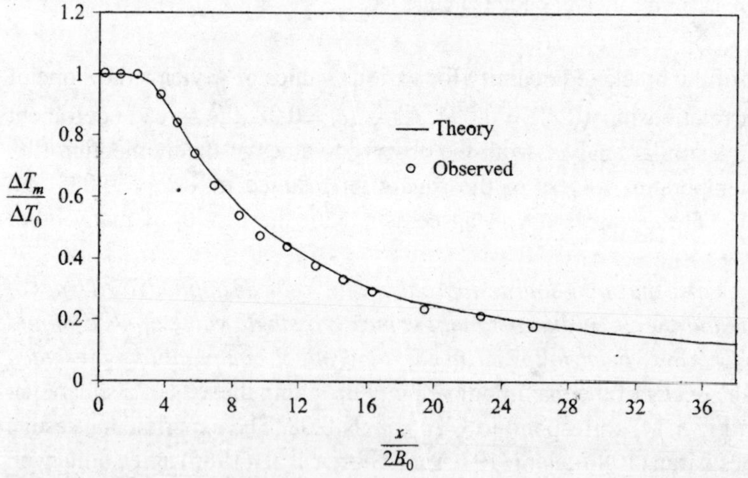

Figure 5.15 Temperature excess decay for a square jet (Source: Pani and Parameswaran, 1994)

zone of a jet. More so, there is no need to apply a correction to account for the location of the virtual origin.

Lateral distribution of momentum flux

In case of three-dimensional jets, the characteristic length scale is $\sqrt{A}$. Therefore, in many of the analysis of the available data, the streamwise distance x has been normalised by $\sqrt{A}$.

The distribution of the momentum flux in the lateral direction of a square jet has been computed using Eq 5.52. For $\frac{x}{\sqrt{A}}$ equal to 5, 10 and 15, a plot of these curves is shown in Fig 5.16. The curves show that for $\frac{x}{\sqrt{A}}$ exceeding 15, the distribution follows closely the self-similar plot of an axially symmetric jet proposed by Maxwell (1970).

The above theoretical features are in agreement with the experimentally observed distribution (Fig 5.17). For $\frac{x}{\sqrt{A}}$ exceeding 10, the distribution of the temperature excess, too, gradually merges with the mean curve proposed by Hinze (1949).

Rectangular outlet of aspect ratio $e = 3$

With an outlet of aspect ratio 3, it is demonstrated that the point-source technique can quite easily be applied to rectangular outlets. The velocity

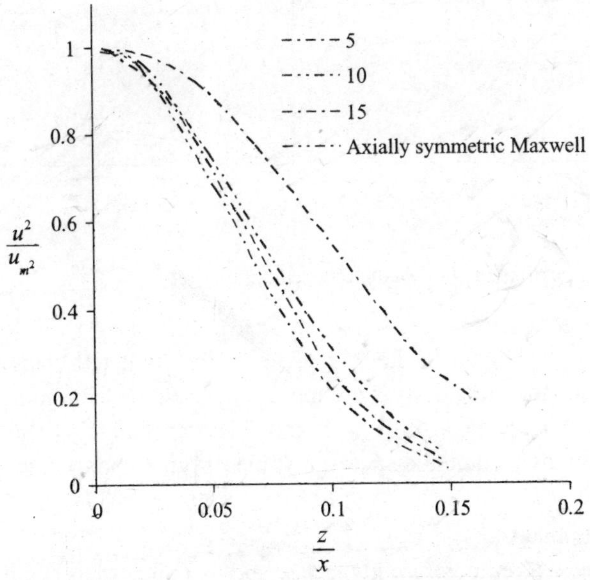

Figure 5.16 Development of lateral momentum flux profiles for a square jet

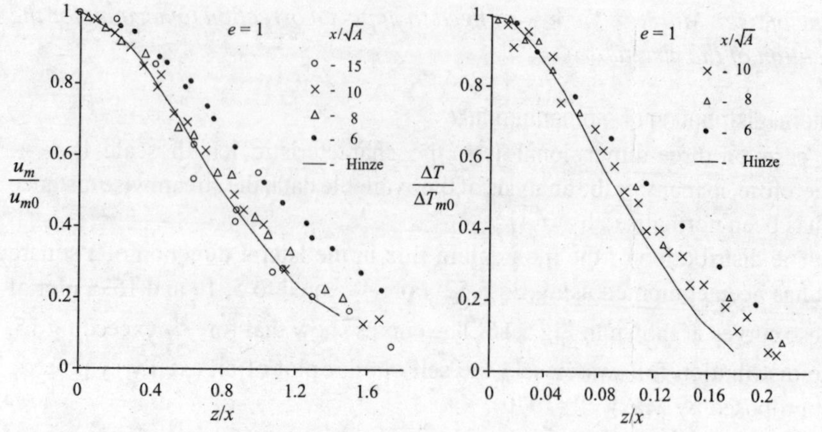

Figure 5.17 Observed distribution of velocity and temperature excess for a square jet

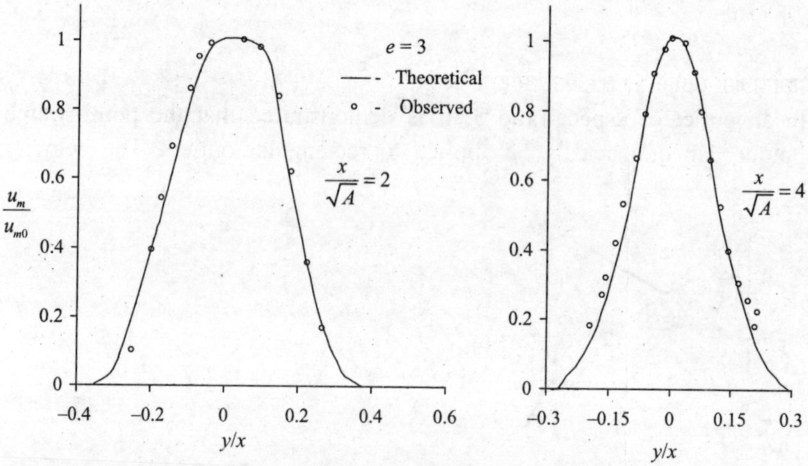

Figure 5.18 Cross velocity profiles in X–Y plane of a rectangular jet

distributions in the x–y planes and in the x–z planes were predicted using the c and λ values meant for the axially symmetric jet. These distributions of the mean velocity are depicted in Figs 5.18 and 5.19, respectively. The agreement between the predicted and the observed values is quite remarkable.

Excess temperature distribution

The distribution of the excess temperature in the x–y and x–z planes has been evaluated for various x-locations, $\frac{x}{\sqrt{A}}$ = 2, 4, 6, 8 and 10. Here again, the

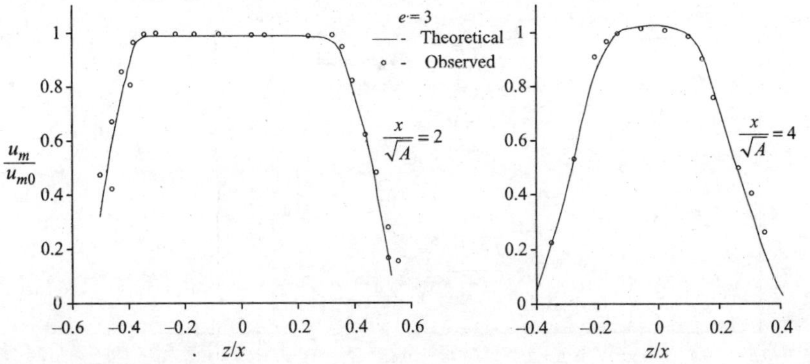

Figure 5.19 Cross velocity profiles in X–Z plane of a rectangular jet (Source: Pani and Parameswaran, 1994)

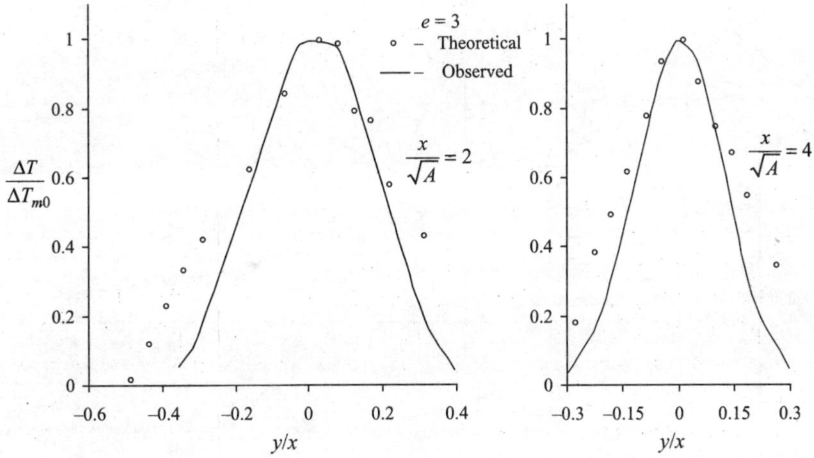

Figure 5.20 Excess temperature profile in X–Y plane of a rectangular jet

agreement between the predicted and the observed values is seen to be good. Typical results for two locations, $\frac{x}{\sqrt{A}} = 2$ and 4, are shown in Figs 5.20 and 5.21.

Decay of u_{m0} and ΔT_{m0}

The decay of the maximum velocity u_{m0} and the maximum temperature excess ΔT_{m0} along the axis of the jet is shown in Fig 5.22. The close agreement between the theoretical and the experimental values is noteworthy.

In contrast to the fore-mentioned approach, Narain (1976) used different values of c for the developing and the developed zones of the three-dimensional jet. Further, different values for c were also selected for the elemental jet, which

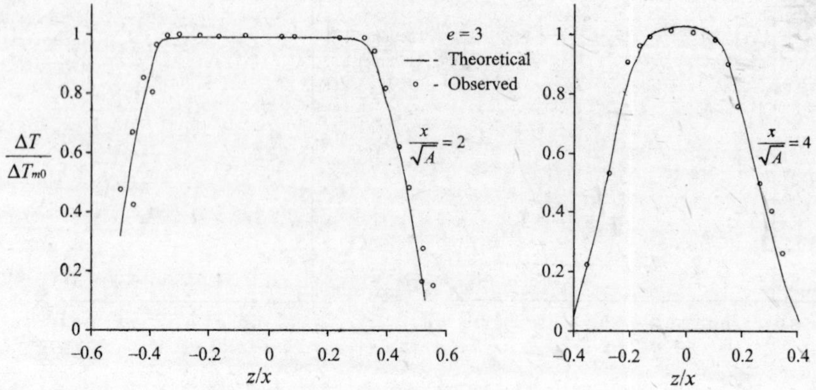

Figure 5.21 Excess temperature profile in X–Z plane of a rectangular jet

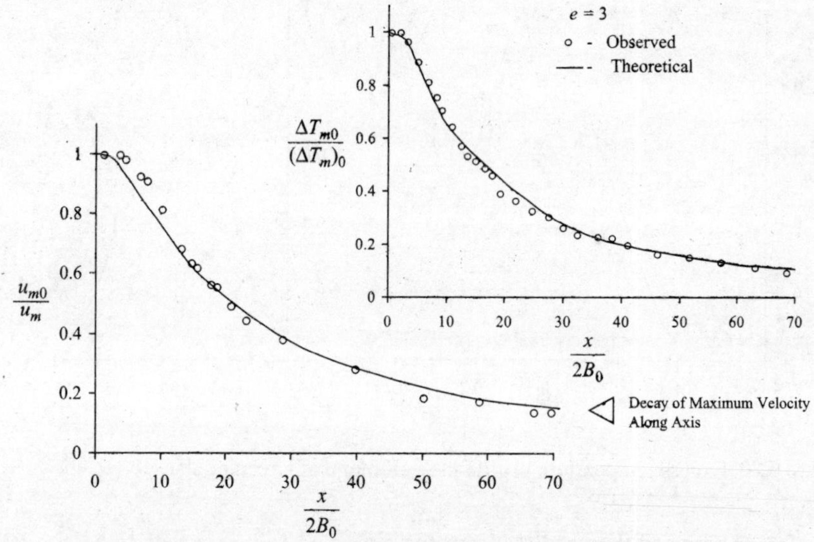

Figure 5.22 Axial velocity and temperature excess decay for a rectangular jet

was the basis for predicting the decay of the maximum velocity of the axially symmetric and the plane jets. *Such an approach of selecting different values for c, on a case-to-case basis, is not in conformity with the basic philosophy of the point-source approach.*

The foregoing approach can only be used to predict the time-averaged velocities, temperatures and the concentration of tracers. The instantaneous values of the above parameters differ quite substantially from the time-mean values depending on the turbulence prevailing in the flow. Therefore, the measurements of the turbulence of three-dimensional jets are presented below. Often for validation, the turbulence properties are used for quite a few predictive models.

5.7.5 Turbulence Measurements

Jets issuing from rectangular nozzles having aspect ratios of 3–10 were used. Measurement of turbulence intensities and the Reynolds stresses was carried out using a constant-temperature hot-wire anemometer (Lozanova and Stankov, 1998). The variation of the turbulence intensity $\frac{\sqrt{u'^2}}{U_0}$ along the axis of the jet is shown in Fig 5.23.

The maximum intensity and its location were found to depend on the initial conditions of the flow. The intensities of the fluctuations in the y and z direction also exhibit a similar behaviour though their maximum values lie within the range 7–12.5 per cent. Nonetheless, all the turbulent fluctuations decay in the downstream direction and tend to attain isotropy.

The variation of the local turbulence intensities $\frac{\sqrt{u'^2}}{u_{m0}}$, $\frac{\sqrt{v'^2}}{u_{m0}}$ and $\frac{\sqrt{w'^2}}{u_{m0}}$ is shown in Fig 5.24. Self-similarity of the fluctuating flow field occurs far

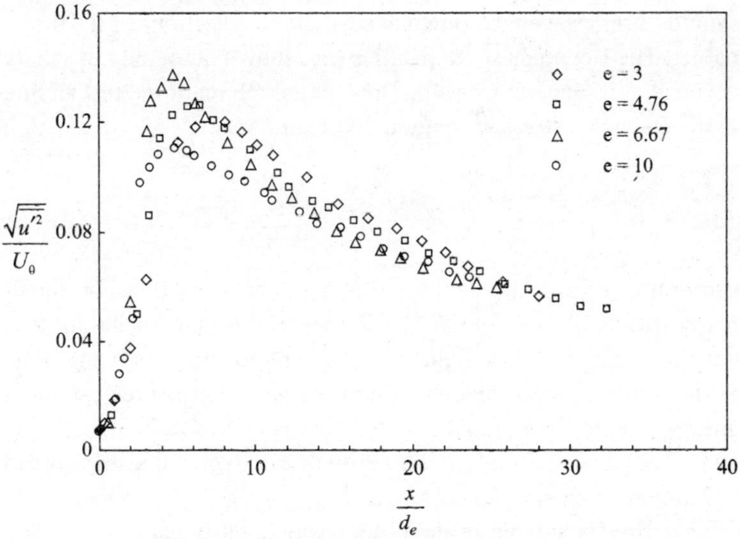

Figure 5.23 Decay of turbulence intensity along the x-axis for jets of different aspect ratios: d_e is the equivalent diameter (Lozanova and Stankov, 1998)

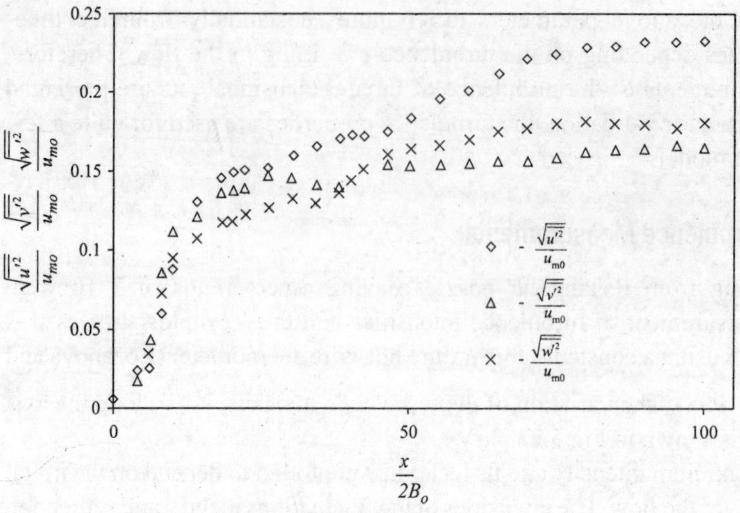

Figure 5.24 Turbulence intensities along the x-axis of a rectangular jet, $e = 10$ (Source: Lozanova and Stankov, 1998)

downstream compared to the similarity of the mean velocity field. In the self-similar flow zone, the value of the longitudinal component of the local turbulence is approximately 24 per cent, whereas the other two components are significantly smaller.

The Reynolds stresses were recorded at several cross sections of the three-dimensional jet. The Reynolds stress profiles (not shown here) did not satisfy similarity even at a distance of $\frac{x}{d_e} = 40$. Thus, in the region of interest arising in practice, the Reynolds stresses do not exhibit similarity.

5.8 Summary

Lateral distribution of the momentum flux and the heat flux for three-dimensional bluff jets has been found by the method of superposition. It is not necessary to divide the flow field into different zones, as is being done by various researchers. For all practical purposes, the point-source can yield a unified mean velocity decay relationship which is valid for the developing and the developed zones of the jet. The growth rate, $\sqrt{2}c$, and the spread coefficient, λ, found from experiments on axially symmetric jets were used to predict the mean flow behaviour of three-dimensional bluff jets.

The decay of the maximum temperature excess, ΔT_{m0}, along the axis of a jet can be predicted from Reichardt's hypothesis. For a bluff jet emanating from any shape of the outlet, $\sqrt{A}$ happens to be the characteristic length scale for normalising the distance x.

The fluctuating flow field exhibits self-similarity farther downstream of the outlet compared to the distance where the mean velocities become self-similar.

Example 5.1: Find the variation of the momentum flux and maximum velocity along the longitudinal axis of a 10-cm-diameter round air jet. Treat the outlet as an equivalent square shape nozzle. Consider the axial distance at which the momentum flux drops to 99 per cent of the value at the efflux section to represent the potential core length, x_0. Compare x_0 with the conventional length of the potential core for a circular jet. Assume $b = 0.0565x$ in the relationship $\frac{u^2}{u_m^2} = e^{-\frac{1}{2}\left(\frac{y}{b}\right)^2}$.

Solution: $d = 10 \text{ cm} = 0.10 \text{ m};\qquad A = \frac{\pi}{4} \times 0.1^2 = 7.854 \times 10^{-3} \text{ m}^2$

$$\sqrt{A} = 2B_0 = 0.08862 \text{ m};\qquad \frac{B_0}{d} = 0.4431$$

To avoid distinction between the different zones of the flow, it will be advantageous to follow the point-source technique. Thus,

$$\frac{u_{m0}^2}{U_0^2} = erf\left(\frac{B_0}{\sqrt{2}b}\right) erf\left(\frac{L_0}{\sqrt{2}b}\right);\qquad \text{for a square nozzle } L_0 = B_0.$$

Hence,

$$\frac{u_{m0}^2}{U_0^2} = \left\{erf\left(\frac{B_0}{\sqrt{2}cx}\right)\right\}^2,\qquad \text{where } c = 0.0565.$$

The variation of the momentum flux and the velocity at various locations along the axis of the square outlet is shown in the following table.

x (cm)	$\frac{x}{d}$	$\frac{B_0}{\sqrt{2}cx}$	$erf\left(\frac{B_0}{\sqrt{2}cx}\right)$	$\frac{u_{m0}^2}{U_0^2}$	$\frac{u_{m0}}{U_0}$	$\frac{u_m}{U_0}\left(=\frac{6}{\frac{x}{d}}\right)$
30	3.0	1.846	0.9909	0.982	0.991	–
60	6.0	0.923	0.8083	0.653	0.808	1.0
100	10.0	0.5538	0.5664	0.321	0.566	0.60
150	15.0	0.3692	0.3980	0.158	0.398	0.40
200	20.0	0.2769	0.3045	0.0927	0.304	0.30
300	30.0	0.1846	0.2059	0.0424	0.206	0.20

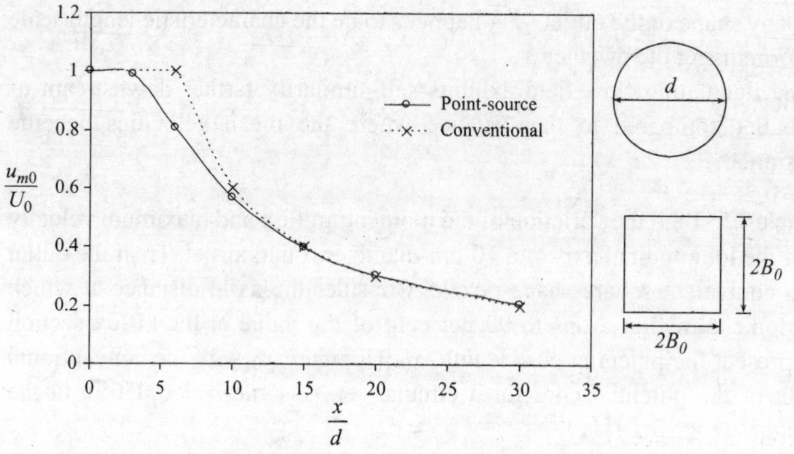

Figure 5.25 Decay of maximum velocity: Example 5.1

To find the length of the potential core, let $\frac{u_{m0}^2}{U_0^2} = 0.99$

Hence, $erf\left(\frac{B_0}{\sqrt{2b}}\right) = \sqrt{0.99} = 0.99498$

Or, from the error function table, $\frac{B_0}{\sqrt{2b}} = 1.985$; Thus, $\frac{B_0}{\sqrt{2cx_0}} = 1.985$

Or, the length of the potential core $\frac{x_0}{B_0} = \frac{1}{1.985 \times \sqrt{2 \times c}} = 6.3$

The same distance can be expressed in terms of the diameter of the nozzle as, $\frac{x_0}{d} = 6.3 \times 0.4431 = 2.8$

Conventionally, a value of $\frac{x_0}{d}$ equal to 6.2 has been accepted in literature on jets. The lacuna is that the value of 6.2 represents the location at which $\frac{u_m}{U_0} = 1$ and the velocity decay curve of the fully developed zone meet.

The variation of the maximum velocity along the longitudinal axis is shown in Fig 5.25. The reason for the large difference in the estimated length of the potential core is due to the way it is defined in the two methods. From the velocity decay plot, it is evident that the observed velocity decay follows the predictions of the point-source technique more closely in the initial regions of the flow.

However, for $\frac{x}{d} > 15$ the conventional method and the point-source method yield almost identical results for the maximum velocity.

Example 5.2: A deeply submerged rectangular outlet 0.5 m deep and 1.0 m wide discharges water at a uniform velocity of 3.0 m/s. Find the maximum velocities on the axis of the jet at distances of 2.0 m and 5.0 m downstream of the efflux section. Also plot the depthwise velocity distribution at $x = 5$ m. Adopt the point-source results.

Solution: To find the maximum velocity on the axis of the jet u_{m0}, we use the relationship, $\frac{u_{m0}}{U_0} = \left\{erf\left(\frac{B_0}{\sqrt{2}b}\right) erf\left(\frac{L_0}{\sqrt{2}b}\right)\right\}^{\frac{1}{2}}$, where $U_0 = 3.0$ m/s, $B_0 = 0.25$ m and $L_0 = 0.5$ m. For the given size of the outlet,

$$\frac{B_0}{\sqrt{2}b} = \frac{B_0}{\sqrt{2}cx} = \frac{0.25}{\sqrt{2} \times 0.0565 \times x} = \frac{3.125}{x},\text{ where } x \text{ is in metres.}$$

Similarly, $\frac{L_0}{\sqrt{2}b} = \frac{0.50}{\sqrt{2}\times 0.0565\times x} = \frac{6.25}{x}$. The maximum velocity at the specified locations is shown below.

x (m)	$erf\left(\frac{3.125}{x}\right)$	$erf\left(\frac{6.25}{x}\right)$	$\frac{u_{m0}}{U_0}$	u_{m0} (m/s)
2.0	0.9729	0.9990	0.9858	2.957
5.0	0.6234	0.9229	0.7585	2.275

In the x–y plane, z being zero, from Eqs 5.52 and 5.53 we have,

$$\frac{u_m^2}{u_{m0}^2} = -\frac{1}{2}\frac{\left\{erf\left(\frac{y-B_0}{\sqrt{2}b}\right) - erf\left(\frac{y+B_0}{\sqrt{2}b}\right)\right\}}{erf\left(\frac{B_0}{\sqrt{2}b}\right)}$$

At $x = 5.0$ m, $\left(\frac{B_0}{\sqrt{2}b}\right) = \frac{0.25}{\sqrt{2}\times 0.0565\times 5.0} = 0.625$, and $erf\,(0.625) = 0.6234$.

Hence, $\frac{u_m^2}{u_{m0}^2} = -\frac{1}{2\times 0.6234}\left\{erf\left(\frac{y-B_0}{\sqrt{2}b}\right) - erf\left(\frac{y+B_0}{\sqrt{2}b}\right)\right\}$

Or, $\frac{u_m^2}{u_{m0}^2} = -0.802\left\{erf\left(\frac{y-0.25}{0.4}\right) - erf\left(\frac{y+0.25}{0.4}\right)\right\}$

From the above expression, the dynamic pressure and the velocity at different y values are computed as shown below.

y (m)	$erf\left(\frac{y-0.25}{0.4}\right)$	$erf\left(\frac{y+0.25}{0.4}\right)$	$\frac{u_m^2}{u_{m0}^2}$	$\frac{u_m}{u_{m0}}$	u_m (m/s)
0.0	−0.6234	0.6234	1.0	1.0	2.275
0.125	−0.342	0.815	0.9279	0.963	2.19
0.25	0.0	0.9229	0.740	0.860	1.956
0.375	0.342	0.9728	0.5059	0.711	1.618
0.50	0.6234	0.992	0.2956	0.544	1.237
0.625	0.8152	0.996	0.1450	0.380	0.866
0.75	0.9229	0.999	0.0610	0.247	0.562

The following figure shows the y distribution of the mean velocity.

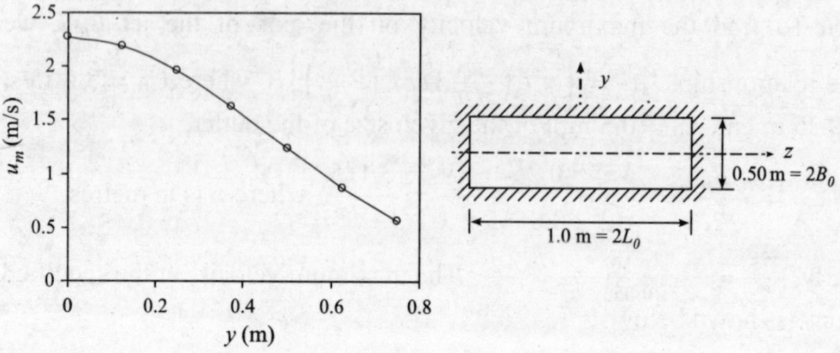

Figure 5.26 Depth-wise distribution of the velocity in the mid-plane; $x = 5.0$ m

6 Coflowing Jets

6.1 Introduction

A coflowing jet occurs when the fluid surrounding a jet is in motion along the axis of the jet. For many practical problems, jets diffuse in a coflowing ambient. They have a wide range of applications, namely, flow from hydraulic outlets in rivers, discharge of wastewater from outfalls into rivers and the sea, jet pumps and ejectors, combustion problems, etc. In case of jets diffusing in a stagnant ambient, the shear flows can be scaled algebraically, that is, b and u_m can be expressed as simple algebraic function of the distance x. However, coflowing jets exhibit different scaling behaviour, which depends on the downstream distance.

The coflowing jet has many common features with the classical jet discharging into a stagnant ambient. These include the formation of a potential core region, followed by the fully developed flow zone, with a transition region in between. In the fully developed region, the excess velocity of a coflowing jet satisfies self-similarity. Therefore, when an observer moves at a constant speed equal to that of the external stream, the velocity profiles at different x cross sections will have the same shape. This feature has been validated experimentally.

A schematic representation of a coflowing jet is shown in Fig 6.1, wherein the excess velocity is defined as $\Delta u = u - U_a$. Let us consider the case when the initial jet velocity, U_0, is significantly higher than U_a. In the near field, that is, for relatively small values of x, the jet will behave more like a pure jet. In this chapter, it will be explained as to how the scaling laws, established earlier for jets in a stagnant ambient, will also be valid for coflowing shear layers. They are as follows:

$$\begin{aligned} \textit{Axisymmetric jet:} \quad & b \propto x, \quad \textit{and} \quad \Delta u_m \propto x^{-1} \\ \textit{Plane jet:} \quad & b \propto x, \quad \textit{and} \quad \Delta u_m \propto x^{-\frac{1}{2}} \end{aligned} \tag{6.1}$$

When the distance x is large, the maximum excess velocity Δu_m becomes small compared to the ambient velocity U_a, and the behaviour of the jet changes

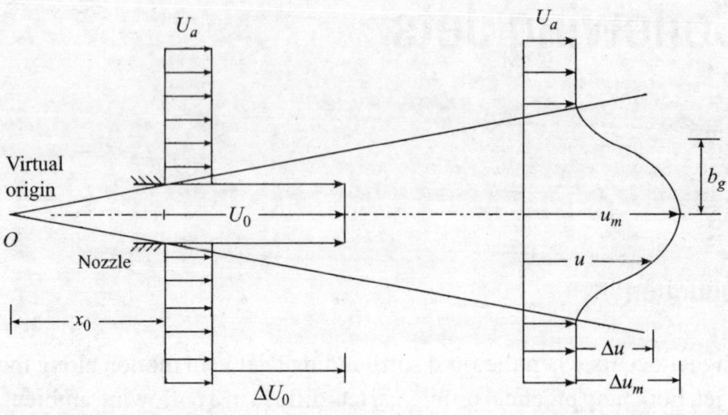

Figure 6.1 Definition sketch of a coflowing jet

significantly. This zone will be designated as the far field and the jet behaves like a weak jet.

An accurate assessment of the dilution of the effluent is often warranted for environmental impact assessment. The longitudinal distances considered in the analysis are fairly large; therefore, such flows from outfalls can be treated as weak jets. The weak jet analysis will predict the dilution of effluents and their adverse impact on the environment.

6.2 Scaling Laws

The principles involved for finding various scaling relationships of a coflowing, axially symmetric jet are presented in brief. Details on the scaling laws are available in the book *Turbulent Jets* (Rajaratnam, 1976).

The equations of motion for an axisymmetric turbulent jet are

$$u\frac{\partial u}{\partial x} + v\frac{\partial u}{\partial r} = \frac{1}{\rho r}\frac{\partial (r\tau)}{\partial r} \tag{6.2}$$

and

$$\frac{\partial (ru)}{\partial x} + \frac{\partial (rv)}{\partial r} = 0 \tag{6.3}$$

Conservation of excess momentum flux

Let us multiply Eq 6.2 by ρr and integrate with respect to r in the limits 0 to ∞. By making use of the continuity equation, Eq 6.3, one can arrive at the following expression,

$$\frac{d}{dx}\int_0^\infty ru(u-U_a)dr = 0 \tag{6.4}$$

Equation 6.4 states that the *excess-momentum* flux is conserved along the direction of flow.

Self-similarity of excess velocity

Let us write the nondimensional distribution of the excess velocity over the cross section of the jet as

$$\frac{(u-U_a)}{(u_m-U_a)} = \frac{\Delta u}{\Delta u_m} = f(\frac{r}{b}) = f(\eta) \tag{6.5}$$

where b is a characteristic length scale. One of the ways of defining b is to assign it a value of r at which $\Delta u = 0.5\Delta u_m$.

First p, q relationship

Based on Eqs 6.4 and 6.5, it can be shown that

$$\frac{d}{dx}\int_0^\infty b^2\eta\left\{\frac{\Delta u_m}{U_a}f + \left(\frac{\Delta u_m}{U_a}\right)^2 f^2\right\}d\eta = 0 \tag{6.6}$$

For a strong jet, $\frac{\Delta u_m}{U_a} >> 1$; hence, Eq 6.6 can be approximated as

$$\frac{d}{dx}\left\{b^2\left(\frac{\Delta u_m}{U_a}\right)^2\right\}\int_0^\infty \eta f^2 d\eta = 0 \tag{6.7}$$

The definite integral in Eq 6.7 will have a finite value, irrespective of the exact nature of the function f. Therefore, we can further simplify the above equation and write

$$\frac{d}{dx}(b^2\Delta u_m^2) = 0 \tag{6.8}$$

If we let, $\Delta u_m \propto x^p$ and $b \propto x^q$, Eq 6.8 leads to

$$2p + 2q = 0 \tag{6.9}$$

In contrast to the above, for a weak jet, $\frac{\Delta u_m}{U_a} \ll 1$. Therefore, from Eq 6.6, we have

$$\frac{d}{dx}\left(b^2\frac{\Delta u_m}{U_a}\right)\int_0^\infty \eta f d\eta = 0 \tag{6.10}$$

$$\text{or} \quad \frac{d}{dx}(b^2 \Delta u_m) = 0$$
$$\text{hence} \quad 2q + p = 0 \tag{6.11}$$

It can also be observed that for the intermediate range, that is, when $\frac{\Delta u_m}{U_a} \sim 1$, the integral momentum equation 6.6 is unable to give any algebraic relationship between p and q.

Second p, q relationship: strong jet

The second relationship between p and q has been formulated based on the properties of self-preserving flows. By considering $u = U_a + \Delta u = U_a + \Delta u_m f$, each term of Eq 6.2 is evaluated one by one. For getting the value of v in terms of Δu_m, η and f, the continuity equation is used. On substitution and simplification of Eq 6.2, the following relationship results,

$$bU_a \frac{\Delta' u_m}{(\Delta u_m)^2} \eta f + \frac{b\Delta' u_m}{\Delta u_m}\left(\eta f^2 - f' \int_0^\eta \eta f \, d\eta\right)$$
$$- \frac{U_a}{\Delta u_m} b' \eta^2 f' - b'\left(\eta^2 f f' - f' \int_0^\eta \eta^2 f' d\eta\right) = g + \eta g' \tag{6.12}$$

We shall assume that the η-dependent terms and the functions f and f' are of the same order of magnitude. It may be noticed that in case of the first and third terms of Eq 6.12, the multiplying factor of the functions involving f and f' is proportional to $\frac{U_a}{\Delta u_m} x^{q-1}$. Similarly, the multiplying factors in the second and fourth terms are proportional to x^{q-1}.

For strong jets, that is, when $\Delta u_m \gg U_a$, the first and third terms of Eq 6.12 can be dropped. As a consequence, for the strong-jet region,

$$\frac{b\Delta' u_m}{\Delta u_m}(\eta f^2 - f' \int_0^\eta \eta f \, d\eta) - b'(\eta^2 f f' - f' \int_0^\eta \eta^2 f' d\eta) = g + \eta g' \tag{6.13}$$

For the flow to be self-preserving, it is essential that $\frac{(b\Delta' u_m)}{\Delta u_m} \propto x^0$, and $b' \propto x^0$. Proportionality condition will be satisfied if $q = 1$. For the strong jet, Eq 6.9 has the relationship $2p + 2q = 0$, which gives a value of $p = -1$. Therefore, in the strong-jet region, the algebraic relations are $\Delta u_m \sim x^{-1}$ and $b \sim x$.

Second p, q relationship: Weak jet

For a weak jet, $\Delta u_m \ll U_a$, therefore, for a weak jet, the first and third terms in Eq 6.12 are dominant. The second and the fourth terms can be neglected.

Retaining only the first and the third terms in Eq 6.12, we have

$$\frac{bU_a\Delta' u_m}{(\Delta u_m)^2}\eta f - \frac{U_a b'}{\Delta u_m}\eta^2 f' = g + \eta g' \tag{6.14}$$

For preservation of self similarity of flow,

$$\frac{b\Delta' u_m}{(\Delta u_m)^2} \propto x^0; \quad and \quad \frac{b'}{\Delta u_m} \propto x^0 \tag{6.15}$$

Both requirements are satisfied if $q - p - 1 = 0$. Earlier, a relationship between p and q was established in Eq 6.11 as $2q + p = 0$. Therefore, the required values of the exponents are $q = \frac{1}{3}$ and $p = -\frac{2}{3}$.

Hence, for a weak jet, we can write, $\Delta u_m \sim x^{-\frac{2}{3}}$ and $b \sim x^{\frac{1}{3}}$. These are identical to the far-wake relationships, albeit, in case of the wake, the velocity defect, $(U_a - u)$, takes the role of the excess velocity of a jet. The mathematical justification for this type of behaviour is that both the far-field wake and the weak jet are governed by the same boundary layer type of equations.

The above approach is also applicable for coflowing plane jets. For a strong plane jet, $p = -\frac{1}{2}$ and $q = 1$.

6.3 Excess Momentum Flux Conservation

Since there is no source or sink of momentum, the excess momentum flux across any plane normal to the x-axis is conserved as stated in Eq 6.4. For axisymmetric jets, the same can be rewritten as follows:

$$M_{e0} = M_e = \int_0^\infty ru(u - U_a)dr = \frac{d^2}{8}U_0(U_0 - U_a) \tag{6.16}$$

where d is the diameter of the outlet, and the efflux velocity U_0 is assumed to be distributed uniformly over the cross-sectional area.

Momentum length scale θ

Let us introduce a momentum length scale θ which satisfies the relationship

$$\theta^2 U_a^2 = M_e \tag{6.17}$$

Therefore, from Eqs 6.16 and 6.17,

$$\theta^2 = \int_0^\infty r\frac{u}{U_a}\left(\frac{u}{U_a} - 1\right)dr = \int_0^\infty \left(1 + \frac{\Delta u}{U_a}\right)\frac{\Delta u}{U_a}rdr \tag{6.18}$$

In the far field $\Delta u \ll U_a$, and the above expression, Eq 6.18, simplifies to

$$\theta^2 = \int_0^\infty \frac{\Delta u}{U_a}rdr \tag{6.19}$$

The excess momentum flux M_e being equal to M_{e0}, from the definition of θ given in Eq 6.17, the momentum length scale remains the same throughout the flow field. If a velocity ratio, $\frac{U_0}{U_a} = m$, is introduced, from Eqs 6.16 and 6.17, we can write $\theta^2 = \frac{d^2}{8} m(m-1)$, for $m > 1$. This relationship also shows that *for a given value of m, the momentum length scale θ does not vary with the distance x.*

The usefulness of the momentum thickness θ as an appropriate parameter for normalising the distance x is evident from the velocity measurements conducted by Biringen (1986) and others. For different values of m, the thickness of the shear layer b and the observed decay of the excess velocity looked very different when plotted against x. However, all the observed values collapsed onto unified curves when x was normalised by θ. Therefore, it is justified to write

$$\frac{b}{\theta} = f_1\left(\frac{x}{\theta}\right) \quad and \quad \frac{\Delta u_m}{U_a} = f_2\left(\frac{x}{\theta}\right) \tag{6.20}$$

Theoretically, b should vary with x in a nonlinear fashion (weak jet case) at very large distances from the efflux section. However, the normalised plots of experimental data reveal that the nonlinear growth of b takes place even after a very short distance. On the contrary, the variation of Δu_m^{-1} with x (strong jet case) follows a linear relationship up to distances of $\frac{x}{\theta} \sim 180$, overriding the theoretical constraint that it should be true only in the initial reach. No convincing explanation has been found for such lacunae. It has been suggested that the range of distances covered in the experiments probably lies in the transition zone between the strong and the weak jet regime. Measurements beyond the said range of distances have been reported recently (Davidson and Wang, 2002).

6.4 Jet Spreading

The general spreading hypothesis (Chu and Lee, 1996) assumes that the change in width of the shear layer in a Lagrangian frame of reference is proportional to the relative velocity between the jet element and its surroundings, that is, $\frac{db}{dt} = c|\Delta u_m|$. In a modified form, we can write the same relationship as

$$\frac{db}{dt} = \frac{db}{dx}\frac{dx}{dt} = u\frac{db}{dx} = (U_a + \Delta u_m)\frac{db}{dx} \tag{6.21}$$

The representative velocity u has been considered as $(U_a + \Delta u_m)$. From Eq 6.21, one can write

$$\frac{db}{dx} = c\frac{\Delta u_m}{(U_a + \Delta u_m)} \tag{6.22}$$

An expression similar to Eq 6.22 was proposed by Abramovich (1963) for the case of an axially symmetric coflowing jet. The physical significance of Eq 6.22 is that, as Δu_m decays with distance x, the growth rate $\frac{db}{dx}$, too, should change along the path of the coflowing jet, resulting in a curved jet boundary as depicted in Fig 6.2.

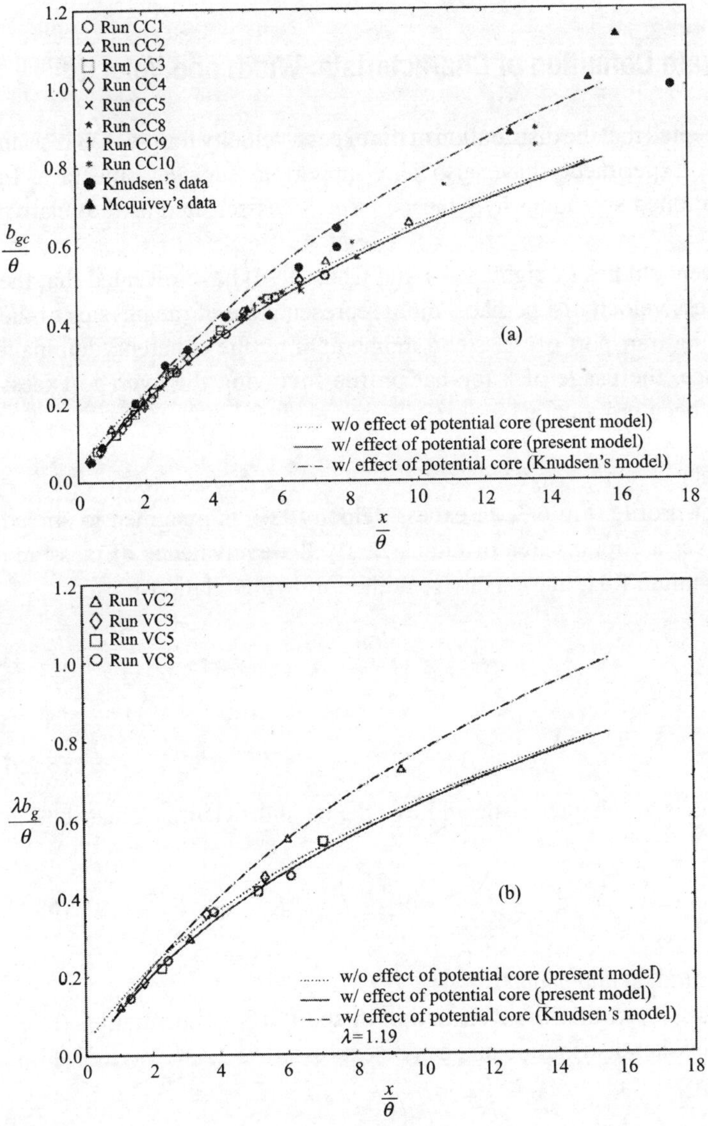

Figure 6.2 Jet boundary in a coflowing stream (Source: Chu, 1996)

It should be noted that the coefficient c appearing in Eq 6.22 represents the rate of growth of a jet in the relationship ($b = cx$) when the ambient is stagnant. Since, in the far field $\Delta u_m \ll U_a$, Eq 6.22 can be simplified as

$$\frac{db}{dx} = c\left(\frac{\Delta u_m}{U_a}\right) \tag{6.23}$$

6.5 Alternate Definition of Characteristic Width and Velocity

It is well accepted that the distribution of the excess-velocity follows a Gaussian distribution. Experiments have also lent support to such a conjecture. In the forementioned spreading hypothesis, $|\Delta u_m|$ was selected as the relative velocity.

Some recent studies (Wright, 1974 and Chu, 1994) have revealed that the average excess velocity is, perhaps, more representative of the physics of the flow involving transport of mass and momentum by the coherent, dominant eddies. Hence, the usage of a top-hat profile involving the average excess velocity has been propounded.

Top-hat profile

For a top-hat profile, Fig 6.3, an excess velocity ΔU is imagined to spread uniformly over a circular area of radius B. By the equivalence of mass and excess momentum flux, one can arrive at the following relationships:

$$\Delta U = \frac{\Delta u_m}{2} \tag{6.24}$$

and

$$B = \sqrt{2}b_g \tag{6.25}$$

In Eq 6.25, b_g is the characteristic width of the jet in the Gaussian distribution

$$\frac{\Delta u}{\Delta u_m} = \mathrm{e}^{-\left(\frac{r^2}{b_g^2}\right)} \tag{6.26}$$

Tracer concentration and dilution

It has been observed that the distribution of the tracer concentration C too follows a Gaussian distribution and can be expressed as

$$\frac{C}{C_m} = \mathrm{e}^{-\left(\frac{r^2}{\lambda^2 b_g^2}\right)} \tag{6.27}$$

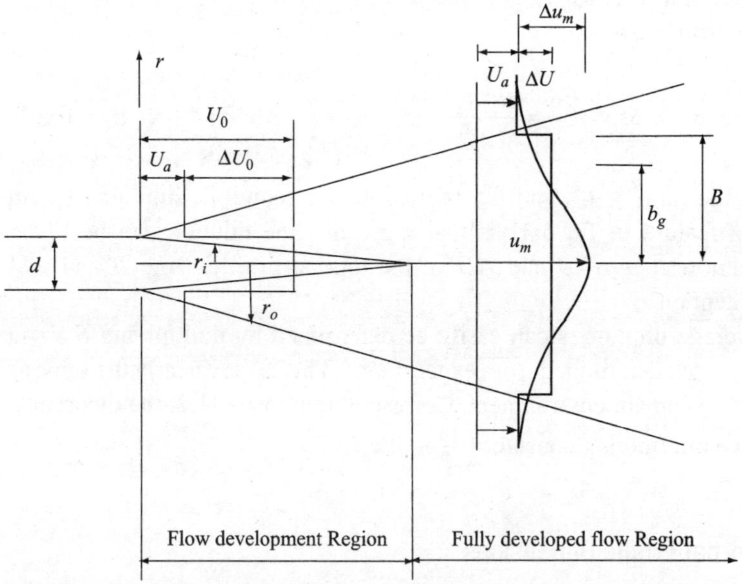

Figure 6.3 Top-hat profile definition sketch

The coefficient λ in Eq 6.27 lies in the range of 1.16–1.20 and points to the fact that the distribution of C is wider than the distribution of the excess velocity.

Further, the mass flux of the tracer at any distance x can be written as

$$M_c = \int_0^{\infty} C(\Delta u + U_a) 2\pi r\, dr \tag{6.28}$$

By substituting for Δu and C from Eqs 6.26 and 6.27, respectively, in Eq 6.28 and integrating it with respect to r, the following relationship can be arrived at:

$$M_c = \pi\lambda^2 b_g^2 C_m \left(U_a + \frac{\Delta u_m}{1+\lambda^2} \right) \tag{6.29}$$

Making use of Eqs 6.24 and 6.25, the tracer flux given by Eq 6.29 can be rewritten in terms of the top-hat profile variables, B and ΔU, as

$$M_c = \frac{\lambda^2 \pi}{2} B^2 C_m \left(U_a + \frac{2}{1+\lambda^2} \Delta U \right) \tag{6.30}$$

Since the tracer mass flux is conserved, $M_c = M_{c0} = C_0 U_0 A = C_0 Q_0$. Therefore, from Eq 6.30,

$$S_c = \frac{C_0}{C_m} = \frac{\lambda^2 \pi B^2}{2Q_0}\left(U_a + \frac{2}{1+\lambda^2}\Delta U\right) \tag{6.31}$$

Note that the ratio of C_0 and C_m represents the centreline dilution, S_c. An interesting feature of Eq 6.31 is that the centreline dilution can be found from the knowledge of B and ΔU of the top-hat profile. Both B and ΔU are dependent on x.

The average dilution $\bar{S}$ can easily be determined by multiplying S_c by a constant ~ 1.66, determined from experiments. This is very nearly the same as for a jet in a stagnant environment. Corresponding to $\lambda = 1.2$, the theoretical value of the multiplying constant $\left(\frac{1+\lambda^2}{\lambda^2}\right) \simeq 1.69$.

6.5.1 Top-hat Profile Dimensions

In this section, the aim will be to find B and ΔU as functions of distance x. From the definition of the momentum thickness and the fact that the excess momentum flux is conserved, the B and Δu relationship can be expressed as

$$\pi B^2(\Delta U + U_a)\Delta U = U_a^2 \theta^2 \tag{6.32}$$

In the non-dimensional form, Eq 6.32 can be written as

$$U_*^2 + U_* - \frac{1}{\pi B_*^2} = 0 \tag{6.33}$$

where $U_* = \frac{\Delta U}{U_a}$ and $B_* = \frac{B}{\theta}$.

Using the average excess velocity in the jet spreading hypothesis, one can express the growth rate of the top-hat as, $\frac{dB}{dx} = c\frac{\Delta U}{U_a + \Delta U}$.

The same spreading rate in a nondimensional form can be rewritten as

$$\frac{dB_*}{dx_*} = c\frac{U_*}{(1+U_*)} \tag{6.34}$$

where $x_* = \frac{x}{\theta}$.

The pair of Eqs 6.33 and 6.34 can be solved numerically for B_* and U_* as an initial value problem using the standard Runge-Kutta method. For this, we need the initial conditions which are set up as follows.

Initial condition

The flow behaves like a strong jet in the initial region; therefore, the initial spreading rate near the exit is expressed as

$$\frac{dB}{dx} = c\frac{\Delta U}{U_0} = c\left(1 - \frac{U_a}{U_0}\right) \simeq \frac{d}{2x_0} \tag{6.35}$$

where x_0 is the distance of the virtual origin located upstream of the nozzle face. As an approximation, the initial condition at the jet outlet can be expressed as $(\Delta U_*)_0 = \frac{\Delta U_0}{U_a}$ and $(B_*)_0 = \frac{d_*}{2}$ for $(x_*)_0 = -\frac{d_*}{2c\left(1-\frac{U_a}{U_0}\right)}$, where $d_* = \frac{d}{\theta}$ and $(x_*)_0 = \frac{x_0}{\theta}$.

It may be recalled that the length scale in the Gaussian velocity distribution $b_g = \left(\frac{B}{\sqrt{2}}\right)$, and the length scale for concentration distribution $b_{gc} = (\lambda b_g)$. These parameters can be found if B is known. Similarly, knowing ΔU of the top-hat profile, the maximum excess velocity $\Delta u_m (= 2\Delta U)$ of the Gaussian distribution can also be evaluated.

6.6 Excess Velocity Distribution

The governing equations of motion for the far wake and the weak jet are identical. Consequently, the theoretical distributions of the velocity defect in a wake flow and the excess velocity of the jet are the same. As the wake has been investigated extensively in the past, the discussions presented here pertain to the wake flow.

The distribution of the velocity defect satisfies self-similarity in the far wake region and can be expressed by the universal law

$$\frac{\Delta u}{\Delta u_m} = f\left(\frac{y}{b}\right) = f(\eta) \tag{6.36}$$

Plane Wake

Schlichting adopted Prandtl's old hypothesis of free turbulence and gave a theoretical distribution for the case of a plane turbulent wake as

$$f(\eta) = (1 - \eta^{\frac{3}{2}})^2 \tag{6.37}$$

Equation 6.37 agrees very closely with the experimental results of Schlichting and Reichardt (Abramovich, 1963).

The distribution of the temperature difference Δt based on Taylor's hypothesis was expressed as

$$\frac{\Delta t}{\Delta t_m} = \sqrt{\frac{\Delta u}{\Delta u_m}} \tag{6.38}$$

Observed dimensionless profiles of the temperature and excess velocity substantiate the relationship given in Eq 6.38 (Abramovich, 1963). The velocity distribution can also be found based on Prandtl's new theory of turbulence, that is, $\varepsilon = \kappa b \Delta u_m$ and is given as

$$\frac{\Delta u}{\Delta u_m} = \mathrm{e}^{-\left(\frac{1}{\varepsilon_0 x}\right) y^2} \tag{6.39}$$

In Eq 6.39, ε_0 is a constant which can be evaluated from experimentally observed velocity profiles. The ordinate b, the point where the velocity defect $\Delta u = 0.367\Delta u_m$, will yield ε_0, since $b = \sqrt{\varepsilon_0 x}$. The difference between the two distributions given by Eq 6.37 and Eq 6.39 is marginal.

Modified Reichardt's equation

In an alternate approach, Reichardt's equation is modified for wake applications. It may be recalled that the original form of the momentum transport equation for a two-dimensional flow is

$$\frac{\partial \overline{u^2}}{\partial x} = \Lambda(x) \frac{\partial \overline{u^2}}{\partial y^2} \tag{6.40}$$

The velocity $u = U_a + \Delta u$, and in the far wake, $\Delta u \ll U_a$; therefore in terms of the velocity defect, Eq 6.40 can be modified as

$$\frac{\partial \Delta u}{\partial x} = \Lambda \frac{\partial^2 \Delta u}{\partial y^2} \tag{6.41}$$

As the first power of Δu is appearing in Eq 6.41, the universal velocity profile as given in Eq 6.36 is adopted in place of a universal excess momentum flux profile. The final solution that can be arrived at (Abramovich, 1963) is as follows

$$\frac{\Delta u}{\Delta u_m} = \mathrm{e}^{-\frac{1}{2}\left(\frac{y}{b}\right)^2} \tag{6.42}$$

Equation 6.42 is the same as given by Eq 6.39; the two b values represent different widths of the same Gaussian distribution.

If there are multiple sources in the coflowing stream, the principle of superposition, however, cannot be effected through Eq 6.42. This is a consequence of the simplifications in which we considered $\Delta u \ll U_a$. The velocity defect appears as a first power in Reichardt's equation as well as in the universal distribution.

Axially symmetric wakes

Using Prandtl's new hypothesis on turbulent shear, the distribution of the velocity defect for the case of an axially symmetric wake is given as

$$\frac{\Delta u}{\Delta u_m} = \mathrm{e}^{-\frac{1}{2\varepsilon_0}\frac{y^2}{x^{\frac{2}{3}}}} \tag{6.43}$$

In the above expression, the constant ε_0 must be determined from experiments. The distributions given in Eqs 6.39 and 6.43 for a plane wake and an axially symmetric wake, respectively, look identical. This is so, since $b \sim x^{\frac{1}{2}}$ for the plane wake, and $b \sim x^{\frac{1}{3}}$ for the axially symmetric wake.

Axially symmetric strong jets

Let us extend Reichardt's hypothesis to the case of axisymmetric coflowing jets by retaining the original form, meant for strong jets. For an elemental area of the source dA, the following holds good:

$$\frac{u(u-U_a)}{U_0(U_0-U_a)} = \frac{dA}{2\pi b^2}\exp\left(-\frac{\xi^2}{2}\right) \tag{6.44}$$

By superposing solutions of elemental areas $dA = r\,dr\,d\phi$, which are spread over the entire cross section of a circular outlet, the excess momentum flux can be written as

$$\frac{u_m(u_m-U_a)}{U_0(U_0-U_a)} = 1-\exp\left\{-\frac{1}{4}\left(\frac{d}{\sqrt{2}cx}\right)^2\right\} \tag{6.45}$$

In terms of the velocity excess, Eq 6.45 can be rewritten in the following form:

$$\frac{u_m}{U_0}\frac{\Delta u_m}{\Delta U_0} = 1-\exp\left\{-\frac{1}{4}\left(\frac{d}{\sqrt{2}cx}\right)^2\right\} \tag{6.46}$$

Let us further denote $\frac{\Delta u_m}{\Delta U_0} = q$ and the ratio $\frac{U_a}{U_0} = \beta$. The left-hand side of Eq 6.46 can be expressed in terms of q and β as

$$\frac{u_m}{U_0}\frac{\Delta u_m}{\Delta U_0} = \beta q + (1-\beta)q^2 \tag{6.47}$$

By combining Eq 6.46 with Eq 6.47, the following relationship, Eq 6.48, can written as

$$\frac{x}{d} = \frac{1}{2\sqrt{2}c}\left\{-ln(1-\beta q-(1-\beta)q^2)\right\}^{-\frac{1}{2}} \tag{6.48}$$

From experimentally observed values of the excess velocity Δu_m at different x distances, the constant $\sqrt{2}c$ can easily be determined using Eq 6.48. For strong jets, including a jet in a stagnant ambient, the coefficient c will have a constant value along the jet trajectory.

Through a combination of Eqs 6.46 and 6.47, the following can be written:

$$a_0 + a_1 q + a_2 q^2 = 0 \tag{6.49}$$

where $-a_0 = 1 - \exp\left\{-\frac{1}{4}\left(\frac{d}{\sqrt{2}cx}\right)^2\right\}$, $a_1 = \beta$ and $a_2 = (1-\beta)$. Knowing the values of β and $\sqrt{2}c$, the value of q at any location x can be evaluated as the positive root of the quadratic equation Eq 6.49.

Non-axisymmetric jets

The above method can be extended to the case of non-axisymmetric outlets as well. For non-axisymmetric outlets, the point source technique (Chapter 5) resulted in distribution of the excess momentum flux as

$$\frac{u_{m0}\Delta u_{m0}}{U_0 \Delta U_0} = erf\left(\frac{B_0}{\sqrt{2}cx}\right) erf\left(\frac{L_0}{\sqrt{2}cx}\right) \tag{6.50}$$

Equation 6.50 can be used to evaluate the coefficient a_0 in Eq 6.49.

For square outlets, $L_0 = B_0$. Without loss of accuracy, square outlets can replace the circular shape, since it is mathematically simpler to handle such a shape. This is true for jets in stagnant as well as in moving ambient, the coflowing jets. The advantages that are inherent in Eqs 6.44 and 6.50 should not be overlooked. In their derivation, the governing equation of motion was linear in terms of $(u\Delta u)$ because of which the solutions can be superposed. Thus, Eq 6.44 or Eq 6.50 can be used to predict the maximum excess velocity decay downstream of an array of rectangular or circular outlets, diffusing in a coflowing stream (Chapter 7).

6.6.1 Generalisation of Reichardt's Hypothesis

For weak jets, the jet boundary being a non-linear function of x, the coefficient c (in the relationship $b = cx$) will vary from section to section, along the trajectory. A method for determining the variation of c with x will be presented later, Eq 6.53.

In this section, Reichardt's hypothesis is being generalised for an axially symmetric coflowing jet, without restricting it to a strong jet. The approach is based on an assumption that *the excess momentum flux satisfies self-similarity and its distribution is Gaussian,*

$$\frac{u\Delta u}{u_m \Delta u_m} = e^{-\frac{\xi^2}{2}} \tag{6.51}$$

where $\xi = \frac{r}{b}$, and b is the radial distance from the jet axis where $u\Delta u$ is 0.606 times $u_m \Delta u_m$.

In the above equation, the product $(u\Delta u) = (U_a + \Delta u)\Delta u$, where U_a has a constant value. Most researchers have found experimentally that Δu satisfies self-similarity and can be described by a Gaussian distribution. However, the important assumption embedded in Eq 6.51, that the distribution of $u\Delta u$ is Gaussian, is not obvious. In the absence of a rigorous proof, an alternate method can be followed to check its validity.

Velocities U_0 and U_a are selected to preset a value of $\beta(= \frac{U_a}{U_0})$. At some distance x, the velocity u_m and Δu_m are predicted based on the available solutions for coflowing jets. Based on a Gaussian distribution of the excess velocity, Δu values are generated. By adding U_a to Δu, all the excess velocity values are modified to have u and u_m. The products $u\Delta u$ can be found and recorded in a tabular form for re-plotting $u\Delta u$ as a function of the radial distance, r. Using the length scale from the new plot, the distribution of $\frac{u\Delta u}{u_m \Delta u_m}$ was plotted as a function of ξ. The steps involved are given in Example 6.2. The computed distribution was also compared with the results available from experiments, and it has been reaffirmed that the distribution of $u\Delta u$ is indeed Gaussian (Pani et al., 2009). Furthermore, following the steps identical to those of the jet in a stagnant ambient, one can arrive at Eqs 6.44 and 6.46.

Link factor

In spite of the practical merits of Reichardt's hypothesis, there is a major constraint on the large eddy spreading hypothesis leading to Eq 6.22. In case of a stagnant ambient, the relationship between b_g and Reichardt's momentum scale b is simple, and $b = 0.5b_g$ for all values of x. This is, however, not true for a coflowing jet. The spreading hypothesis for a coflowing jet needs to be modified and expressed as

$$\frac{db}{dx} = k_l c_g \left\{ \frac{\frac{\Delta u_m}{U_a}}{2 + \frac{\Delta u}{U_a}} \right\} \tag{6.52}$$

In the above equation, k_l is called a link factor, and c_g $(= 0.114)$ is the spread rate of the velocity profile in a stagnant ambient. The link factor involving Δu_m

is given by the following:

$$k_l = \sqrt{\frac{\left(\frac{\Delta u_m}{U_a}\right)+2}{4\left(\frac{\Delta u_m}{U_a}+1\right)}} \tag{6.53}$$

The proof of the above expression, Eq 6.53, is given in Example 7.2. It is based on the principle of conservation of the excess momentum. The coefficient k_l changes along the jet trajectory as the ratio $\frac{\Delta u_m}{U_a}$ becomes smaller with increasing distance, x. In case of a strong jet, including the case of a stagnant ambient, $k_l \simeq 0.5$. For a weak jet, $\frac{\Delta u_m}{U_a} \to 0$ and $k_l \simeq 0.707$. In the transition zone, the link factor assumes values which lie between the two above mentioned limits.

The two unknowns, Δu_m and b, have been predicted by solving numerically the excess momentum conservation equation and the modified spreading hypothesis. As expected, these results were identical to the values computed by adopting the top-hat profile. The advantages of Reichardt's hypothesis in dealing with a single coflowing jet is not very obvious. Nonetheless, in case of multiple coflowing jets, it provides a sound theoretical basis for superposing the excess momentum flux $u\Delta u$ resulting from individual jets. This aspect will be dealt with in Chapter 7.

6.7 Experimental Results

The outcome of important experiments conducted by various investigators on coflowing jets is presented in brief. It is natural to find some difference in their results because of the different test conditions and the measuring techniques adopted by them. Often, it is difficult to make a comparative study as it is hard to access the original data.

6.7.1 Circular Jets

The results of Landis and Shapiro (Abramovich, 1963) are used to plot the excess momentum flux at various streamwise locations in the form given by Eq 6.48. From the slope of the resulting linear plot, the values of the spread coefficient $\sqrt{2}c$ were found as 0.08(0.08), 0.0476(0.048) and 0.0327(0.028) for different β such as 0.0, 0.25 and 0.46, respectively. The quantities appearing in brackets are the values predicted from Abramovich's empirical formula, $\sqrt{2}c = (\sqrt{2}c)_0 \frac{1-\beta}{1+\beta}$.

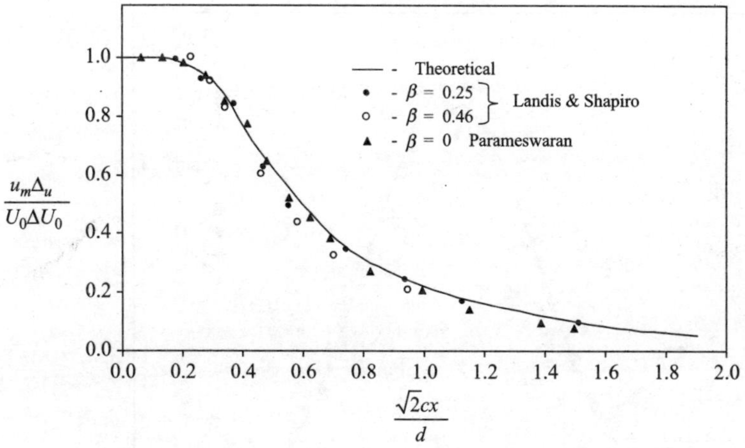

Figure 6.4 Decay of excess momentum along axis of a coflowing round jet

Based on another set of temperature and velocity measurements in a stagnant ambient, the values of $(\sqrt{2}c)_0 = 0.08$ and the coefficient of spread $\lambda = 1.17$ were derived (Parameswaran, 1986). However, making use of LIF and LDA measuring techniques, it was found that $(\sqrt{2}c)_0 = 0.076$ and $\lambda = 1.2$ (Chu, 1996).

To cross check the accuracy of $\sqrt{2}c$, the excess momentum flux $\frac{\Delta u_m}{\Delta U_0}\frac{u_m}{U_0}$ is plotted as a function of the non-dimensional distance $\sqrt{2}c\frac{x}{d}$ for different values of β, as shown in Fig 6.4. The agreement between observed data and the theoretical curve, described by Eq 6.46, looks reasonable. This is not surprising since the $\sqrt{2}c$ values were determined using the same set of data.

Concentration profiles have been measured by many investigators, and are found to follow the Gaussian distribution. As an example, the distribution of concentration and velocity measured by Chu (1996) with LIF and LDA techniques, respectively, are shown in Fig 6.5.

To get a comprehensive knowledge about the concentration and excess velocity distributions, one needs to know the scales C_m, b_{gc}, Δu_m and b_g as a function of x. The wake theory does indicate the way Δu_m and b_y should vary with x in the near and the far field. However, where exactly the changes take place have been found from experiments. The momentum thickness θ serves as an appropriate parameter for normalising x. The parameter θ can be expressed in terms of known parameters, d and β, as $\theta = \frac{d}{2}\sqrt{\frac{1}{\beta}\left(\frac{1}{\beta} - 1\right)}$. In practice, the actual values of the distance lie mostly in the transition zone, and the theory is not of much help.

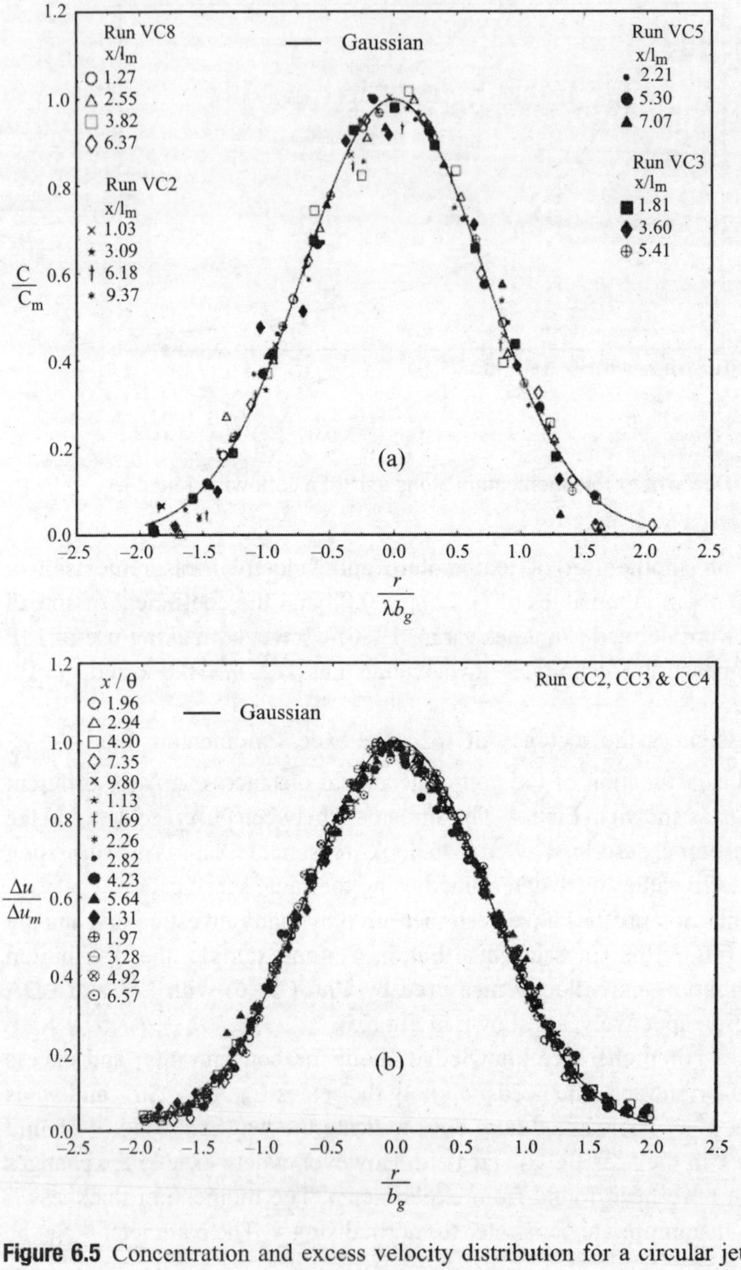

Figure 6.5 Concentration and excess velocity distribution for a circular jet (Source: Chu, 1996)

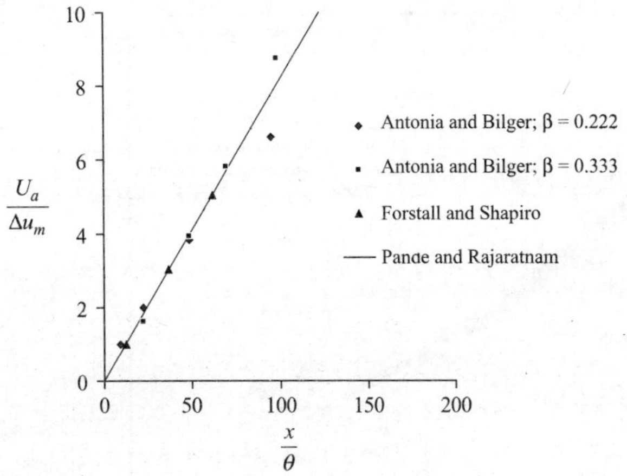

Figure 6.6 Variation of maximum excess velocity with distance (Source: Rajaratnam, 1976)

Using the integral momentum equation and the integral energy equation, it is possible to express $\frac{U_a}{\Delta u_m}$ as a function of $\frac{x}{\theta}$, shown in Fig 6.6 (Pande and Rajaratnam, 1975). The plot includes the experimental data from other sources and the predicted values from Forstall and Shapiro's empirical relationship

$$\frac{\Delta u_m}{\Delta U_0} = \frac{4 + 12\beta}{\left(\frac{x}{d}\right)} \tag{6.54}$$

As the agreement between the predicted and the observed data seems to be good, Eq 6.54 could be used to find the velocity scale. Alternatively, from Fig 6.6, one could get

$$\frac{\Delta u_m}{U_a} = \frac{12.5}{\left(\frac{x}{\theta}\right)} \tag{6.55}$$

From Chu's plot shown in Fig 6.7, $\frac{\Delta u_m}{U_a} = \frac{7}{\frac{x}{\theta}}$ can be fitted to the data points. Observed values of other investigators seem to corroborate the relationship $\frac{\Delta u_m}{U_a} = \frac{7}{\frac{x}{\theta}}$. The significant difference in the predictions from the two plots, Figs 6.6 and 6.7, is due to the fact that θ has been defined differently in each. By multiplying the numerator of Chu's relationship by a factor of $\sqrt{\pi}$, one can get parity between the two relationships. With a knowledge of $\frac{\Delta U}{U_a}$ (or $\frac{\Delta u_m}{U_a}$), the length scale B (or b_g) can be evaluated from the conservation of excess momentum flux, such as Eq 6.32.

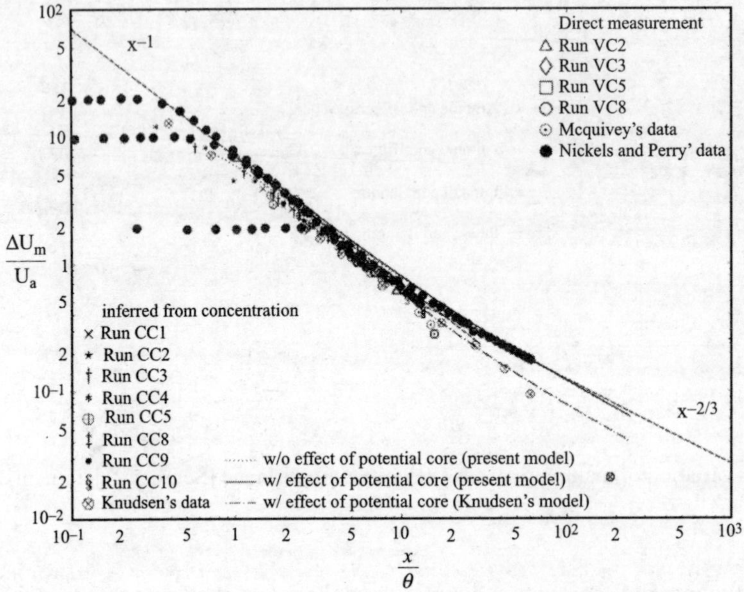

Figure 6.7 Variation of maximum excess velocity with distance (Source: Chu, 1996)

Dilution

The centreline dilution given by Eq 6.31 has been computed for $\lambda = 1.2$ and compared with experimental observations, as shown in Fig 6.8. To make the plot more general, one can use dimensional analysis for expressing the centreline dilution as

$$S_c(x)\left(\frac{\beta}{1-\beta}\right) \propto \left(\frac{x}{\theta}\right) \tag{6.56}$$

The measured values give a linear relationship between the two non-dimensional parameters of Eq 6.56. The constant of proportionality is found to be 0.174.

For predicting the concentration field, the length scale $b_{gc} = \lambda b_g$ can be obtained by assuming $\lambda = 1.2$. Alternatively, b_{gc} can be read from Fig 6.2, since the predicted values have been confirmed by experiments. For predicting the excess temperature distribution and the decay of the maximum excess temperature ΔT_m, a procedure identical to finding the concentration can be followed.

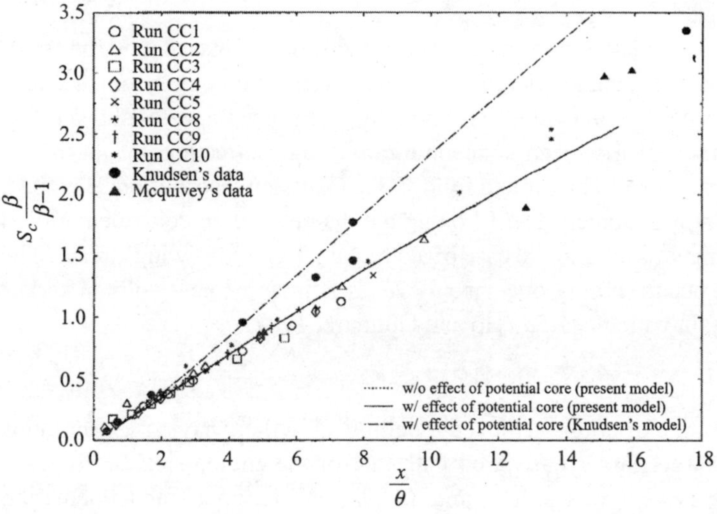

Figure 6.8 Variation of the centreline dilution (Source: Chu, 1996)

Weak jet

In a recent investigation, the range of the experiments was extended to $\frac{x}{d} \sim 1200$ to capture the characteristic properties of the weak-jet flow (Davidson and Wang, 2002). In the far field, represented by $\frac{x}{\theta} = 215$, 536 and 1207, the distribution of $\frac{C}{C_m}$ with $\frac{r}{b_g}$ was observed to exhibit self-similarity and was Gaussian. A consolidated plot of their own and other's data, showed the way b_{gc} varied with the distance $\frac{x}{\theta}$ in the near field, as well as in the far field. For $\frac{x}{\theta} < 4$, a linear relationship exists as predicted for the case of a strong jet. A transition zone persists from $\frac{x}{\theta} \approx 4$ to $\frac{x}{\theta} \approx 100$. Beyond the transition zone, the length scale $b_{gc} \sim x^{\frac{1}{3}}$ is in accordance with the weak-jet (or the far-wake) theory. However, the decay of the maximum concentration was linear up to $\frac{x}{\theta} \sim 100$, and is followed by a transition zone. Only for $\frac{x}{\theta} > 300$ does the $\frac{2}{3}$ power decay law meant for weak jets seems to hold good. The data on the velocity spread, b_g, also showed a similar behaviour. The value of λ was noticed to increase in the weak-jet region by about 30 per cent.

6.7.2 Non-axisymmetric Outlets

Experimental observations on coflowing rectangular and elliptical shaped jets are limited.

Axis-switching

The theoretical relationship, given in Eq 6.50, does not account for 'axis-switching', a phenomenon that has been observed for non-axisymmetric jets. Experiments of Hakem et al. (2007) on elliptical-shaped nozzles showed that the distribution of the mean concentration C along directions of the minor and major axis is self-similar and follows the Gaussian distribution. Beyond a distance of approximately $13d$ (d being the diameter of an equivalent circle) the jet mixing is similar to the case of a circular jet in a coflowing stream. The axis-switch occurred only once at $x \sim 7d$. It is in agreement with an earlier observation, in which $x \sim 5d$ (Ho and Gutmark, 1987).

Dilution

Interestingly, the dilutions achieved by elliptical outlets (having aspect ratios of 3 and 6) were larger compared with that of the circular jet. In the tests, the velocity ratio β was 0.0285 and 0.0192, and the mean and fluctuating components of the concentration were recorded using the LIF technique. The best-fit lines of the experimental data for the centreline dilution can be expressed as

$$\begin{aligned} \textit{Elliptical}: \quad & S_c \simeq 6.5\frac{x}{\theta} \\ \textit{Circular}: \quad & S_c \simeq 4.8\frac{x}{\theta} \end{aligned} \tag{6.57}$$

The expressions of Eq 6.57 are based on observations covering distances up to $\frac{x}{\theta} \sim 1.1$, which are in the strong-jet region. The above expression for the circular jet shows the relative enhanced performance of an elliptical jet. The effect of β is not taken into account. The constant of proportionality, 4.8, is also quite different from the one used in the generalised expression of Eq 6.56.

Investigations by Ho and Gutmark (1987) showed that the elliptical jets having a small aspect ratio of 2:1 entrained the ambient fluid by as much as 3 to 8 times more. Therefore, the dilution should be higher by a similar factor. Even, Mebarka et al. (2007) noticed that jets having larger aspect ratios can enhance dilution, though to a lesser extent. The cause for higher entrainment has been attributed to axis-switching and an increase in the level of turbulence in the flow. Experiments on coflowing rectangular outlets, per se, are scarce. However, a geometry somewhat similar to a rectangular opening, the duck bill valve (DBV), was tested in a coflowing stream (Karindikar, 1997). The DBV was lip-shaped, 15 mm in width, and the thickness of the opening fluctuated between 1 to 3 mm depending on the flow. The aspect ratio varied from 15 to 5. In the analysis, the momentum length scale θ was

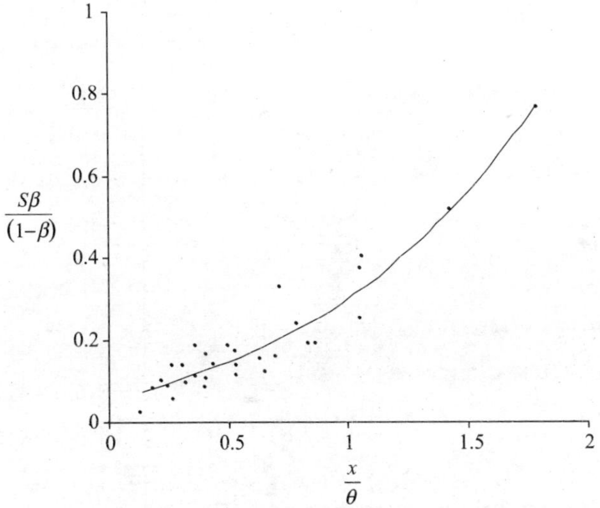

Figure 6.9 Dilution of tracer for a duck bill valve

defined in terms of the excess velocity in the pipe instead of using the excess-velocity at the lip of the valve. Such a modification from the conventional definition of θ has been justified as it provides an equal basis for comparing the performance of DBV and the circular outlet. For circular outlets, their diameters are normally equal to the diameter of the pipe. It has been observed that the DBV enhances mixing, leading to greater dilution of the tracer. This is depicted in Fig 6.9.

6.7.3 Turbulence Characteristics

The radial variation of the concentration intermittency factor γ for a round jet, observed by Chu (1996), in a coflowing ambient is shown in Fig 6.10. The value of γ is nearly equal to 1 in the mid core region of the jet, that is, for $\frac{r}{b_{gc}} <\approx 0.6$, and falls to a negligible value for $\frac{r}{b_{gc}} > 1.70$. With an increase in the distance from the outlet, the width of the core tends to shrink. This means, more of the surrounding fluid is able to enter the core as the size of the eddy grows in the downstream direction. Furthermore, the edge of the top-hat profile is found to lie in the close proximity of the $\gamma = 0.5$ contour.

The rms values of the concentration fluctuations were observed to increase in a steep fashion beyond $\frac{x}{\theta} \sim 100$. Simultaneously, the value of the centreline intermittency factor, γ, dropped from a value of nearly 1 at $\frac{x}{\theta} \approx 100$ to about 0.6 for $\frac{x}{\theta} \approx 1000$ (Davidson and Wang, 2002). These features again indicate

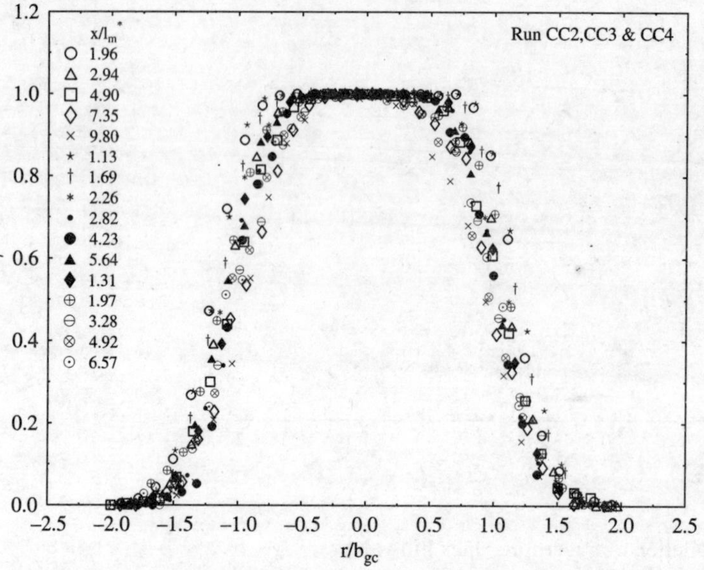

Figure 6.10 Intermittency factor for a round jet (Source: Chu, 1996)

a change in the structure of the turbulent eddy leading to greater intrusion of the ambient field into the central region of the flow.

Figure 6.11 shows the profiles of the mean velocity fluctuations $\sqrt{u'^2}$, $\sqrt{v'^2}$ and $\sqrt{w'^2}$. The magnitudes of lateral and transverse fluctuating components in Fig 6.11 are approximately the same. In the beginning, both are somewhat smaller compared with the fluctuations in the x-direction, but at larger distances, they tend to attain isotropy. The self-similarity of these distributions is evident. It is important to note that self-similarity of the fluctuating components and the turbulent shear stress occurs at much larger values of $\frac{x}{d}$ as compared with the distance needed for the mean velocity profiles to become self-similar. Figure 6.12 shows the distribution of the shear stress $\overline{u'v'}$.

6.8 Summary

The laws governing the decay of the maximum excess velocity and the spread rate of coflowing jets are nearly identical to the variation of the velocity defect and spread of wakes. In the strong-jet region, $\Delta u_m \propto x^{-1}$ and $b_g \propto x$, and in the far field (i.e. the weak-jet region), $\Delta u_m \propto x^{-\frac{2}{3}}$ and $b_g \propto x^{\frac{1}{3}}$. Experimental results corroborate these relationships. The lateral distribution of the tracer

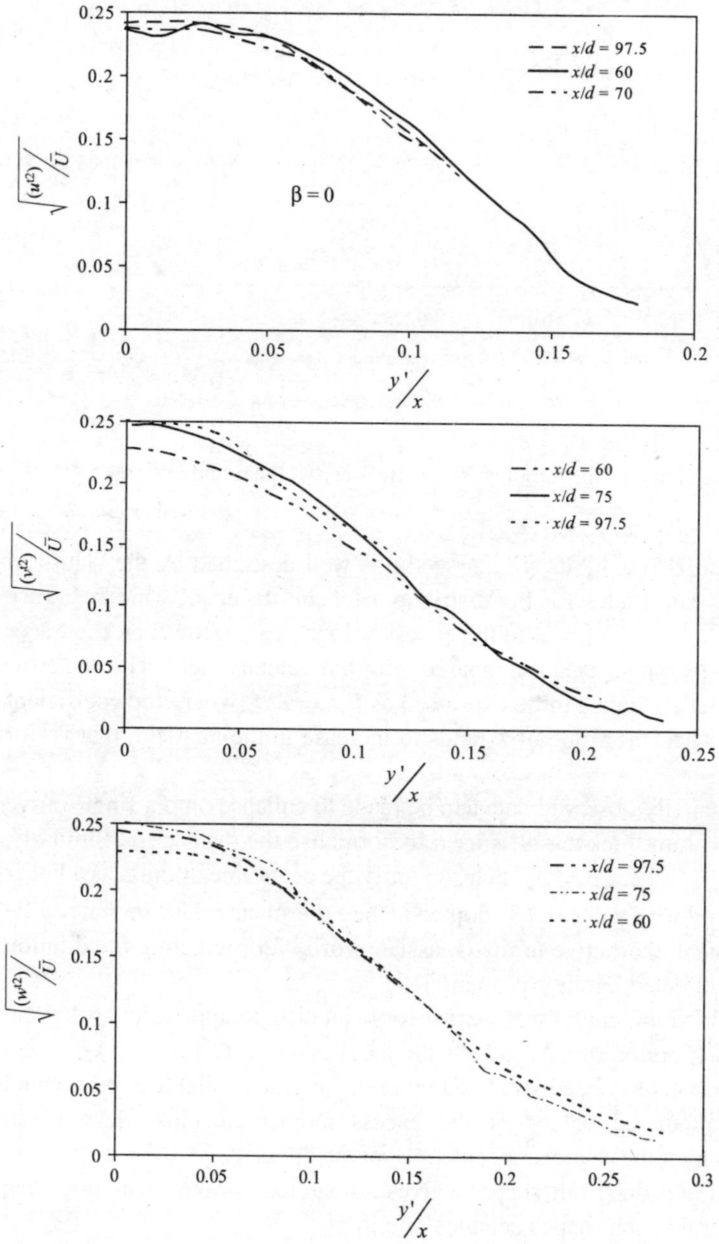

Figure 6.11 Intensity of velocity fluctuations across the round jet (Wygnanski and Fiedler, 1969)

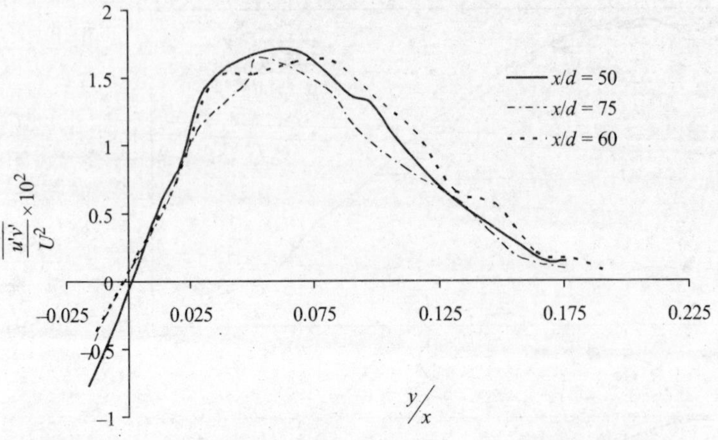

Figure 6.12 Variation of the turbulent shear stress across the round jet

concentration also exhibits similarity and is well described by the Gaussian distribution. The scales for the distribution of the tracer, C_m and b_{gc}, vary with x in a manner identical to the excess velocity case. However, the tracer is observed to spread faster compared with the velocity field. The scale for the concentration field can be expressed as $b_{gc} = \lambda b_g$, where the coefficient $\lambda = 1.2$. Further, the value of λ tends to increase in the far field, that is, for $\frac{x}{\theta} >\approx 100$.

Experimentally observed data can be made to collapse onto a single curve if the momentum thickness θ is used to normalise the distance x. Similarly, the factor $\frac{\beta}{1+\beta}$, where $\beta = \frac{U_a}{U_0}$, helps to unify the centreline dilution as a linear function of $\frac{x}{\theta}$ for all values of β adopted in the experiments. The top-hat profile is a convenient alternative to the Gaussian profile for predicting the dilution that can be effected by the coflowing jet.

Reichardt's momentum transport theory can also be applied to coflowing jets. The distribution of $u\Delta u$ across the jet is assumed to follow a Gaussian distribution. This has been validated by revisiting the available experimental data. The predicted values of the excess momentum flux, $u\Delta u$, from Reichardt's hypothesis are amenable to superposition.

Elliptical and duck-bill shaped valves are subject to axis-switching. Jets emanating from such shapes enhance dilution.

Example 6.1: Ignoring the effects of the molecular viscosity, for a zero pressure gradient, incompressible, axially symmetric jet the momentum equation in terms of the instantaneous velocities, U and V takes the form

$$\frac{\partial(\overline{U^2})}{\partial x} + \frac{1}{r}\frac{\partial(r\overline{UV})}{\partial r} = 0 \tag{6.58}$$

In the presence of a coflowing stream, having a velocity of U_a, the momentum equation can be rewritten as follows:

$$\frac{\partial\left\{\overline{U(U-U_a)}\right\}}{\partial x} + \frac{1}{r}\frac{\partial\left\{r\overline{V(U-U_a)}\right\}}{\partial r} = 0 \tag{6.59}$$

Reichardts hypothesis is modified to express the rate of excess momentum transfer in the transverse direction. Consequently, it looks logical to write

$$\rho\overline{(U-U_a)V} = -\Lambda\frac{\partial}{\partial r}\left\{\rho\overline{(U-U_a)U}\right\} \tag{6.60}$$

Furthermore, the distribution of the excess momentum flux, $\rho\overline{(U-U_a)U}$ may be assumed to satisfy similarity, that is,

$$\frac{\rho\overline{(U-U_a)U}}{\rho\overline{(U_m-U_a)U_m}} = f(\xi) \tag{6.61}$$

where $\xi = \frac{r}{b}$.

Neglecting the contributions of the fluctuating velocity components, derive an expression for the distribution of the excess momentum flux in terms of the mean velocities u and u_m.

Solution: For convenience, we shall write $(U - U_a) = \Delta U$, and ρ being a constant, the momentum equation can be written as

$$\frac{\partial}{\partial x}(\overline{U\Delta U}) + \frac{1}{r}\frac{\partial}{\partial r}(r\overline{V\Delta U}) = 0 \tag{6.62}$$

Introducing Reichardt's relationship, $\overline{\Delta UV} = -\Lambda\frac{\partial}{\partial r}(\overline{U\Delta U})$, the momentum equation can be written as follows:

$$\frac{\partial}{\partial x}(\overline{U\Delta U}) + \frac{1}{r}\frac{\partial}{\partial r}\left\{-\Lambda r\frac{\partial}{\partial r}(\overline{U\Delta U})\right\} = 0 \tag{6.63}$$

For a strong jet, the maximum excess velocity $\Delta U_m >> U_a$. Hence, $\Delta U_m \simeq U_m$. Based on the way the maximum velocity and the width of the jet vary with the distance x, it is reasonable to assume the excess momentum flux to vary inversely as the square of the characteristic thickness. Thus, $\overline{U_m \Delta U_m} = \frac{m}{b^2}$, where m is the constant of proportionality.

From the principle of similarity, $\overline{U\Delta U} = \overline{U_m \Delta U_m} f(\xi) = \frac{m}{b^2} f(\xi)$. Using this relationship, let us evaluate each term of the equation of motion, Eq 6.63.

$$\frac{\partial}{\partial x}\left(\overline{U\Delta U}\right) = \frac{\partial}{\partial x}\left(\frac{m}{b^2} f\right) = -2\frac{m}{b^3}\frac{db}{dx} f + \frac{m}{b^2} f' \frac{\partial \xi}{\partial x} = -\frac{m}{b^3}\frac{db}{dx}\{\xi f' + 2f\}$$

Similarly,

$$\frac{\partial}{\partial r}(\overline{U\Delta U}) = \frac{\partial}{\partial r}\left\{\frac{m}{b^2} f\right\} = \frac{m}{b^2}\frac{\partial f}{\partial r} = \frac{m}{b^2}\frac{\partial \xi}{\partial r} f' = \frac{m}{b^3} f'$$

Also, $\frac{\partial^2}{\partial r^2}\left(\overline{U\Delta U}\right) = \frac{\partial}{\partial r}\left(\frac{mf'}{b^3}\right) = \frac{m}{b^3} f'' \frac{\partial \xi}{\partial r} = \frac{m}{b^4} f''$.

Substituting the above terms in the equation of motion and on simplification, we will have

$$-(2f + \xi f') = \frac{\Lambda}{b}\frac{dx}{db}\left\{f'' + \frac{f'}{\xi}\right\} \tag{6.64}$$

In order to make the dynamic pressure profiles universal, let us set $\frac{\Lambda}{b}\frac{dx}{db}$ a constant. Without loss of generality, the constant can be selected equal to unity. Thus, the mixing length parameter, $\Lambda(x) = b\frac{db}{dx}$. The ordinary differential equation in f, Eq 6.64, simplifies to

$$2f + \left(\xi + \frac{1}{\xi}\right) f' + f'' = 0 \tag{6.65}$$

The solution of the above equation which satisfies the boundary conditions, $f = 1$ for $\xi \to 0$ and $f = 0$ for $\xi \to \infty$ is

$$f(\xi) = \mathrm{e}^{-\frac{\xi^2}{2}} \tag{6.66}$$

The above solution is exceedingly important and will be used in the next chapter to predict the velocity excess downstream of coflowing multiple jets based on the principle of superposition.

Example 6.2: Experiments show that in case of a round jet, the excess velocity, Δu, follows a normal distribution $\frac{\Delta u}{\Delta u_m} = \mathrm{e}^{-\left(\frac{r^2}{b_g^2}\right)}$, where b_g is the radial distance at which $\Delta u = 0.3678 \times \Delta u_m$. In a stagnant ambient, $b_g = 0.109 \times x$.

Consider an air jet, circular in shape, having a diameter 10 cm, discharging into a coflowing stream of air. The efflux velocity $U_0 = 75$ m/s, and the velocity of the coflowing stream $U_a = 15$ m/s. Find the following,

(a) The momentum thickness θ. Will it be the same at different x locations?
(b) The maximum velocity u_m at $x = 1.0$, 2.5 and 5.0 metres downstream of the nozzle.
(c) The radial distribution of Δu at $x = 2.5$ m.
(d) The distribution of $\frac{u\Delta u}{u_m \Delta u_m}$ versus $\frac{r}{b}$, where b is that value of r at which $u\Delta u = 0.606 \times (u_m \Delta u_m)$.

Use the distribution of (c) for generating the excess momentum distribution in (d).

Comment on the final results.

Solution:

(a) Given, $U_0 = 75.0$ m/s, $d = 0.10$ m, $U_a = 15.0$ m/s
The excess momentum $M_{e0} = M_e = A_0 U_0 \Delta U_0 = \frac{\pi}{4} \times 0.1^2 \times 75 \times 60 = 35.34\ m^4/s$.
Also, by definition, $M_e = \theta^2 U_a^2$. Thus, $\theta = \sqrt{\frac{35.34}{15^2}} = 0.396$ m. The value of θ will be the same at all values of x as the excess momentum flux is conserved.
(b) It may be recalled that the plot, Fig 6.7 represents the solution of Eqs 6.33 and 6.34. In a given problem, it is desirable to solve the two simultaneous equations, numerically. However, for the present case, we shall read the values of $\frac{\Delta u_m}{U_a}$ from Fig 6.7, and compute Δu_m and u_m as shown in the table.

x, m	$\frac{x}{\theta}$	$\frac{\Delta u_m}{U_a}$	Δu_m, m/s	u_m, m/s
1.0	2.52	2.8	42.0	67.0
2.5	6.31	1.3	19.5	34.5
5.0	12.61	0.6	9.0	24.0

(c) For finding the excess velocity distribution in the transverse direction at $x = 2.5$ m, the width b_g will be needed. From Fig 6.2, $\frac{b_{gc}}{\theta} = 0.49$, corresponding to $\frac{x}{\theta} = 6.31$.
Hence, $b_g = \frac{0.49 \times \theta}{\lambda} = \frac{0.49 \times 0.396}{1.2} = 0.1617$ m. Using this value of b_g and with $\Delta u_m = 19.5$ m/s, the entries in columns 2 to 4 of the table below are based on the distribution $\frac{\Delta u}{\Delta u_m} = e^{-\left(\frac{r}{b_g}\right)^2}$.

(d) Columns 5 and 6 are generated making use of the known u_m and Δu_m at $x =$ 2.0 m. From a plot of $\frac{u\,\Delta u}{u_m\,\Delta u_m}$ versus r, shown in figure 1 below, the value of b to be used in Reichardt's theory is found to be equal to 0.093 m. Column 7 gives the values of the normalised distance $\frac{r}{b}$ corresponding to the values of r in column 1. The values shown in the columns 6 and 7 are plotted in figure 2.

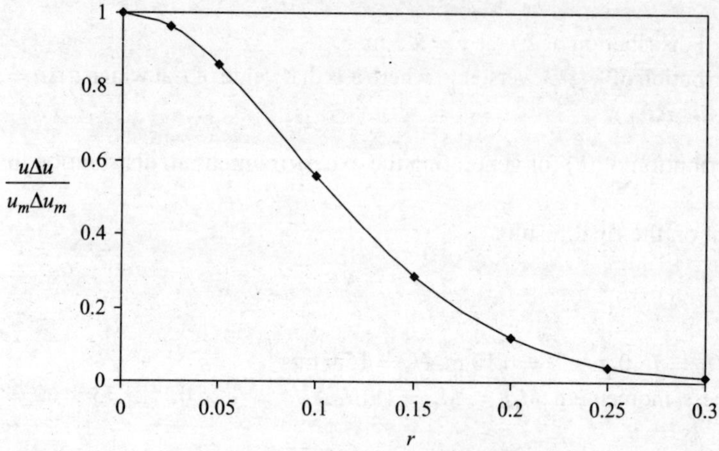

Figure 1 Variation of the excess momentum flux across the round jet

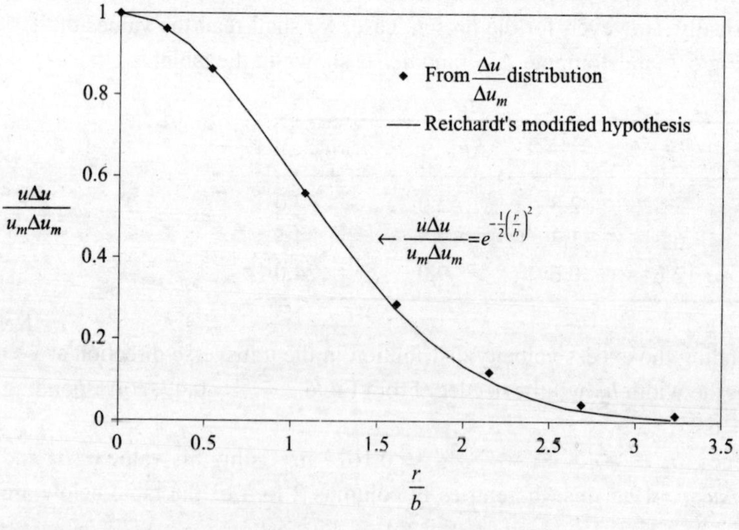

Figure 2 Normalised plot of the excess momentum flux

Comments: It is interesting to note that the distribution of the excess momentum flux agrees very closely with the theoretical distribution that evolved from the modified Reichardt's hypothesis.

r (m)	$\frac{r}{b_g}$	$\frac{\Delta u}{\Delta u_m}$	Δu	u	$\frac{u\Delta u}{u_m \Delta u_m}$	$\frac{r}{b}$
0	0	1.0	19.50	34.5	1.0	0
0.025	0.154	0.976	19.03	34.03	0.962	0.269
0.05	0.309	0.909	17.72	32.72	0.862	0.537
0.10	0.618	0.682	13.30	28.30	0.559	1.075
0.15	0.925	0.425	8.29	23.29	0.287	1.613
0.20	1.237	0.216	4.21	19.21	0.120	2.15
0.25	1.546	0.086	1.68	16.68	0.042	2.69
0.30	1.855	0.032	0.62	15.62	0.014	3.22

Example 6.3: A rectangular outlet, 30 cm deep and 60 cm wide, discharges water into a natural water course at an uniform velocity of 3.0 m/s. The coflowing water course has a speed of 0.6 m/s and being fairly deep, the effects of the boundaries could be ignored and the flow treated as a coflowing free jet. Find the following,

(a) $\frac{\Delta u_{m0}}{\Delta U_0}$ and $\frac{u_{m0}}{U_0}$ at a distance of 10.0 m downstream of the efflux section.
(b) Approximate total width of the shear layer.
(c) Distribution of $\frac{u\Delta u}{u_{m0}\Delta u_{m0}}$ along the y-axis at $x = 10$ m.

Solution: Certain approximations are necessary as information on coflowing rectangular outlets is not readily available. The given outlet may be imagined to be composed of two square outlets, 30 cm × 30 cm each, placed side by side. The equivalent diameter of a circular outlet = 33.85 cm.

For a stagnant ambient, the coefficient of spread of the jet in the point-source method $c \simeq 0.054$. In a coflowing stream, as c is a variable, the local value at $x = 10$ m needs to be evaluated first.

(a) The excess momentum flux of the square outlet, $M_{oe} = M_e = (AU_0)\Delta U_0 = (0.3 \times 0.3) \times 3.0 \times (3.0 - 0.6) = 0.648\ \text{m}^4/\text{s}^2$

As $U_a^2\theta^2 = M_e$, $\theta = \sqrt{\frac{0.648}{0.6^2}} = 1.341$ m

$\frac{x}{\theta} = \frac{10.0}{1.341} = 7.45$

From Fig 6.2, $\frac{b_{gc}}{\theta} = 0.62$. Hence, $b_g = \frac{0.62 \times 1.341}{1.2} = 0.693$ m, and for Reichardt's distribution, $b = \frac{b_g}{2} = 0.346$ m.

With this knowledge of b, that is, cx, from Eq 6.50, $\frac{u_{mo}\Delta u_{mo}}{U_0 \Delta U_0} = erf\left(\frac{0.15}{\sqrt{2}\times 0.346}\right) erf\left(\frac{0.30}{\sqrt{2}\times 0.346}\right) = 0.3353 \times 0.6140 = 0.2059$.
Let us set, $-a_0 = 0.2059$, $a_1 = \beta = \frac{U_a}{U_0} = \frac{0.60}{3.0} = 0.20$, and $a_2 = 1 - \beta = 1 - 0.2 = 0.8$.
From Eq 6.49, we can write, $-0.2059 + 0.2q + 0.8q^2 = 0$, where $q = \frac{\Delta u_{mo}}{\Delta U_0}$. The positive root of the quadratic equation yields $q = 0.397$. Hence, $\Delta u_{mo} = 0.397 \times \Delta U_0 = 0.397 \times 2.4 = 0.954$ m/s. The maximum velocity $u_m = U_a + 0.954 = 1.554$ m/s. Thus, $\frac{u_{mo}}{U_0} = \frac{1.544}{3.0} = 0.518$.

It may be noted that an equivalent circular outlet was adopted to find the value of b_{gc}, since the width of the coflowing jet is not sensitive to either the shape, or because of the presence of the adjecent jet. However, while evaluating the excess velocity, Eq 6.50 meant for a rectangular outlet is appropriate. Usage of Fig 6.7 will be erroneous and is not recommended for rectangular outlets.

(b) The total width of the shear layer may be considered as approximately 4.0 times b_g. Thus, $b_g \approx 4.0 \times 0.693 = 2.77$ m.

(c) The distribution of the excess momentum along the y-axis can be found, since we have the local value for cx, that is, b. This will be identical to the procedure enumerated in Example 5.2

Example 6.4: A 0.3 m diameter, deeply submerged, single port outfall discharges effluent into a coflowing stream. The discharge velocity $U_0 = 3.0$ m/s and the stream velocity $U_a = 0.5$ m/s. The initial concentration of the tracer, $c_0 = 500$ ppm. Evaluate at intervals of 5 m the approximate variation of the jet width, the excess velocity and the average dilution for the first 25 m.

Solution: Given, $d = 0.3$ m; hence, $A = \frac{\pi}{4} \times 0.3^2 = 0.0707$ m^2

Excess momentum, $M_{0e} = M_e = A \times U_0 \times \Delta U_0 = 0.0707 \times 3.0 \times 2.5 = 0.530$ m^4/s^2

The momentum length scale, θ, can be evaluated from the relationship, $\theta^2 U_a^2 = M_e$. Thus, $\theta = \sqrt{\frac{0.530}{0.5^2}} = 1.456$ m.

In general, the excess velocity and the width of the jet can be found from the simultaneous solution of the Eqs 6.33 and 6.34. Figures 6.2 and 6.7 represent the solution in a graphical form. To minimise the inaccuracies while reading the values of the length scale, b_{gc}, from the plot, it will be convenient to fit a polynomial that can follow the actual solution as close as possible. The variation of $\frac{b_{gc}}{\theta}$ with the normalised distance $\frac{x}{\theta}$ can be expressed as,

$$\frac{b_{gc}}{\theta} = 0.01707 + 0.10325\left(\frac{x}{\theta}\right) - 0.00554\left(\frac{x}{\theta}\right)^2 + 1.46585 \times 10^{-4}\left(\frac{x}{\theta}\right)^3$$

For a specified value of $\frac{x}{\theta}$, one can evaluate the width of the jet, b_{gc} (or B) from the above expression. Subsequently, Eq 6.33 can be solved for the velocity-excess. The steps are shown in a tabular form for $x = 5.0$ m, and other rows in the table are a repetition of the same procedure. One should be cautious not to extrapolate the polynomial beyond the range of the fit, that is, $\frac{x}{\theta} > 18$, as inaccuracies may creep in.

x(m)	$\frac{x}{\theta}$	$\frac{b_{gc}}{\theta}$	B^*	$\frac{\Delta U}{U_a}$	$\frac{\Delta u_m}{U_a}$	Q_0	S_c	S_{av}
5	3.434	0.3122	0.3680	1.1126	2.2252	0.2121	2.926	4.86
10	6.868	0.5123	0.6038	0.5596	1.1193	0.2121	6.011	9.98
15	10.302	0.6530	0.7696	0.3873	0.7746	0.2121	8.820	14.64
20	13.736	0.7699	0.9074	0.2978	0.5957	0.2121	11.578	19.22
25	17.170	0.8986	1.0590	0.2306	0.4612	0.2121	15.073	25.02

Illustration: For $x = 5$ m, $\frac{x}{\theta} = \frac{5.0}{1.456} = 3.434$

From the fitted polynomial, the corresponding value of $\frac{b_{gc}}{\theta} = 0.3122$.

Assuming $\lambda = 1.2$, $B^* = \frac{B}{\theta} = \frac{\frac{b_{gc}}{\theta} \times \sqrt{2}}{1.2} = 0.3680$.

With this known value of B^*, Eq 6.33 is solved for U^*, that is, $\frac{\Delta U}{U_a}$, to arrive at the value of 1.1126.

Further, $\frac{\Delta u_m}{U_a} = 2 \times \frac{\Delta U}{U_a}$. Thus, $\frac{\Delta u_m}{U_a} = 2 \times 1.1126 = 2.2252$ (column 6).

The discharge $Q_0 = A_0 \times U_0 = 0.0707 \times 3.0 = 0.2121\ \text{m}^3/\text{s}$. Hence, S_c, the minimum dilution which occurs on the centreline of the jet can be evaluated

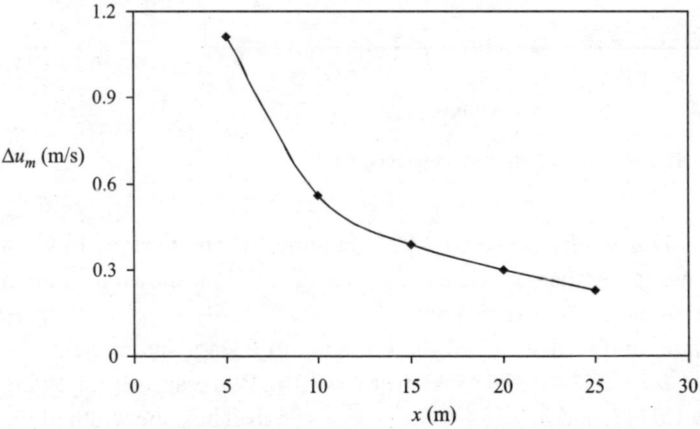

Figure 3 Variation of the excess velocity along the axis of the jet

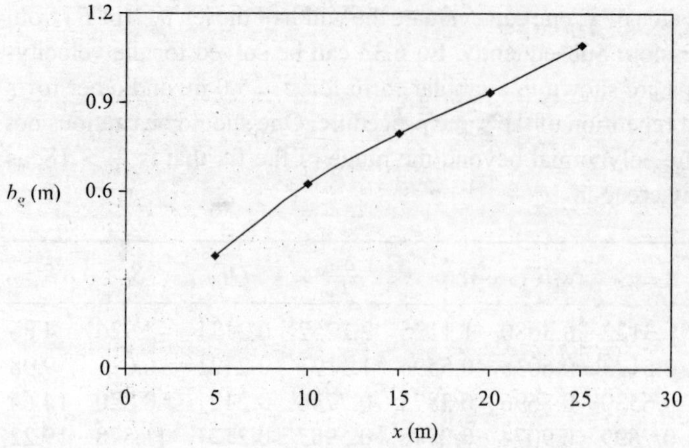

Figure 4 Variation of the width of the jet with x

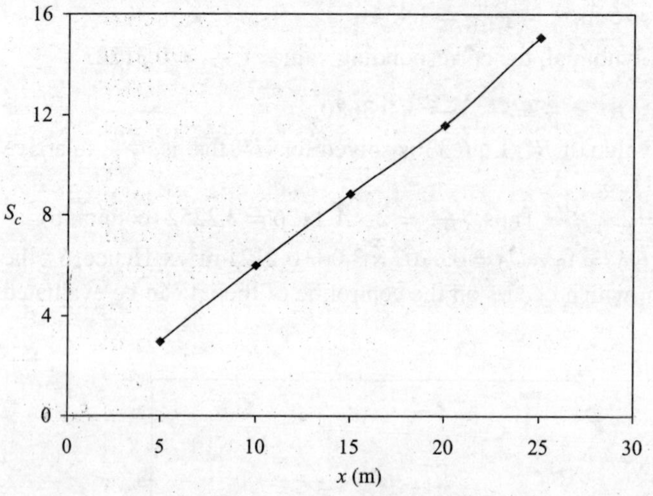

Figure 5 Centreline dilution along the jet trajectory

from Eq 6.31. This works out to be 2.926 (column 8). The average dilution S_{av} is 1.66 times larger than S_c. Hence, the average dilution shown in column 9 is equal to 2.928×1.66, that is, 4.86.

Had the same outfall discharged the effluents to a stagnant ambient, the width of the jet, $b_g = 0.114 \times 30 = 3.42$ m at $x = 30$ m. Whereas, in a coflowing stream, $b_{gc} = 1.0747$, and $b_g = \frac{1.0747 \times 1.456}{\sqrt{2}} = 1.106$ m. Thus, the width of the jet is significantly small in a coflowing environment.

Similarly, the average dilution that can be attained in a stagnant ambient equals $0.29 M_0^{0.5} x Q_0^{-1} = 0.29 \times (0.0707 \times 3^2)^{0.5} \times 30.0 \times \ (0.0707 \times 3.0)^{-1} = 32.7$.

It is interesting to note that, even though the width of the jet is quite small in a coflowing stream, the average dilution that can be achieved is marginally superior compared to the stagnant ambient value. From the graphical plots, Figures 3–5, it is clear that the excess velocity Δu_m drops rapidly at the beginning of the journey, and the jet expands in a non-linear fashion as it moves downstream. The centreline dilution is nearly a linear function of the distance x.

7 Multiple Free Jets

7.1 Introduction

Multiple free jets can mix effluents and other releases more efficiently with the surrounding fluid. For example, sea outfalls mostly have multiple ports to achieve dilution of the effluent as quickly as possible upon its release. Multiple jets are also common in combustion applications, gas burners of power plants, noise control, aerospace industry etc. Flow downstream of multiple jets, arranged in an array or a circular formation, is more complex compared to that of a single jet described in Chapter 6.

In the initial sections of this chapter, the case of multiple jets in a stagnant environment has been considered. It will be followed by the diffusion of an array of multiple jets in a coflowing stream. The theoretical formulations are based on Reichardt's theory, which allows superposition of the excess momentum flux. In the development and analysis of the theoretical expressions, a square shaped outlet is sometimes considered in lieu of the circular orifice. Without loss of accuracy, such a step simplifies the computations of the flow field.

7.2 Stagnant Ambient

7.2.1 Jets from an Array of Outlets

The flow field downstream of a series of round and rectangular holes has been studied by many like Knystautas (1964), Fabis and Fejer (1974), Dash (1981), Parameswaran (1986), Hogdson et al., (1999), and Panayotis et al., (2006). Knystautas found the square of the velocity profiles downstream of an array of circular nozzles, based on Reichardt's theory. A similar procedure was adopted by Dash to predict the momentum flux distribution for the case of multiple three-dimensional jets. Hogdson extended the work to study the concentration distribution downstream of multiple round jets. The recent work of Panayotis

(2006) includes two different techniques: the method of superposition and the method of restricted area. However, *the method adopted by almost all of the investigators has one major limitation, that is, it only holds good in the zone of established flow.*

Merger of jets

For large distances downstream of the nozzles, the flow field gets transformed to that of a plane jet. Thus, the jets diffuse in isolation in the beginning, followed by the merger of the adjacent jets and finally the flow behaves like a plane jet. Recent work on coalescing axisymmetric turbulent plumes (Kaye and Linden, 2004) shows that the merging process depends upon the restricted entrainment that occurs due to the presence of a nearby plume. As a result, the actual merging height is significantly smaller than the height at which isolated plumes are expected to merge. However, this aspect will not be of much consequence for the far field diffusion of jets. Therefore, in our presentation, we will ignore this aspect.

Distribution of momentum flux and tracers: single jet

The theoretical predictions based on Reichardt's hypothesis agree quite well with the experiments, even to the extent of predicting the mean velocity profiles before attaining two-dimensionality. In all cases, the total momentum flux is found to be very nearly conserved. The variation of the maximum momentum flux along the direction of flow and the distribution of the square of the velocities across the jet were established theoretically (see Chapter 4) as

$$\frac{u_m^2}{U_0^2} = \frac{A}{2\pi b^2} \tag{7.1}$$

and

$$\frac{u^2}{u_m^2} = \exp\left(-\frac{\xi^2}{2}\right) \tag{7.2}$$

where $\xi = \frac{r}{b}$.

For scalar quantities like concentration of a tracer, C (or excess temperature ΔT), Eqs 7.1 and 7.2 can be modified and expressed as

$$\frac{u_m C_m}{U_0 C_0} = \frac{A}{2\pi b_c^2} \tag{7.3}$$

and

$$\frac{uC}{u_m C_m} = \exp\left(-\frac{\zeta^2}{2}\right) \tag{7.4}$$

where $\zeta = \frac{r}{b_c}$. For details, refer to section 4.3.

Making use of the point-source technique, the distribution of the specific momentum flux u^2, and the variation of its maximum value along the x-axis of a rectangular outlet of size $(2B_0 \times 2L_0)$ have been expressed as (derivation given in section 5.5)

$$\frac{u^2}{U_0^2} = \frac{1}{4}\left\{erf\left(\frac{y-B_0}{\sqrt{2}b}\right) - erf\left(\frac{y+B_0}{\sqrt{2}b}\right)\right\} \times \left\{erf\left(\frac{z-L_0}{\sqrt{2}b}\right) - erf\left(\frac{z+L_0}{\sqrt{2}b}\right)\right\} \tag{7.5}$$

and

$$\frac{u_{m0}^2}{U_0^2} = erf\left(\frac{B_0}{\sqrt{2}b}\right) erf\left(\frac{L_0}{\sqrt{2}b}\right) \tag{7.6}$$

For a square nozzle, $B_0 = L_0$. Therefore,

$$\frac{u_{m0}^2}{U_0^2} = \left\{erf\left(\frac{B_0}{\sqrt{2}b}\right)\right\}^2 \tag{7.7}$$

In Eq 7.7, b is the characteristic length scale in the below mentioned expression

$$\frac{u_m^2}{u_{m0}^2} = e^{-\frac{1}{2}\left(\frac{r}{b}\right)^2} \tag{7.8}$$

and is equal to the distance from the axis at which u_m^2 is 0.606 times u_{m0}^2.

Dual mixing-length approach

The dual mixing length approach has been applied to account for the differential rate of growth of the shear layer in the y and z directions of three-dimensional outlets. In such cases, the normalised specific momentum flux is written as

$$\frac{u^2}{u_{m0}^2} = \exp\left\{-1.385\left(\eta_y^2 + \eta_z^2\right)\right\} \tag{7.9}$$

where $\eta_y = \frac{y}{b_y}$ and $\eta_z = \frac{z}{b_z}$, and the length scales $b_y = (b_{0.5})_y$ and $b_z = (b_{0.5})_z$. The length scales, $(b_{0.5})_y$ and $(b_0.5)_z$ represent the half widths. They are the distances along y and z at which $u_m = 0.5u_{m0}$ and $u = 0.5u_m$, respectively. Based on the principle of conservation of the forward momentum flux, the variation of the maximum momentum flux on the axis ($y = 0$ and $z = 0$) is given by

$$\frac{u_{m0}^2}{U_0^2} = 0.44\frac{A}{b_y b_z} \tag{7.10}$$

For an individual jet, either Eq 7.6, or Eq 7.10 can be used to find $\frac{u_{m0}^2}{U_0^2}$, prior to applying the principle of superposition of u^2 values of each jet in the array. The circular jet is treated as a specific case of the three-dimensional jet, with $b_y = b_z$ and $A = \frac{\pi}{4}d^2$ in Eqs 7.9 and 7.10. If required, for simplicity of computations, the circular outlet can also be replaced by an equivalent square nozzle of size $2B_0 = 0.886d$.

Selection of a single 'c' value

The jet width 'b' in Eq 7.6 is found from the relationship $b = cx$. The value of c is determined from experiments on circular nozzles. The coefficient c should be independent of the geometry and the aspect ratio of the outlets under consideration. Heuristically, c should be the same for the plane jet as well. Therefore, the equation is not subject to any specific restriction. It can be applied to the developing flow region of jets. Expressions Eq 7.1, and Eq 7.6 yield identical values for the momentum flux distribution in the developed flow region. For the reasons propounded in favour of the point source, *Eq 7.6 gives a unified solution and has numerable advantages,* which have been elaborated throughout the book.

7.2.2 Predictions for Multiple Outlets

Figure 7.1 is a schematic layout of 'N' number of jets, with a uniform spacing of 's' along the z-axis. The location of the virtual origin, x_0, and the distance at which the flow becomes effectively two-dimensional, x_{2D}, will primarily

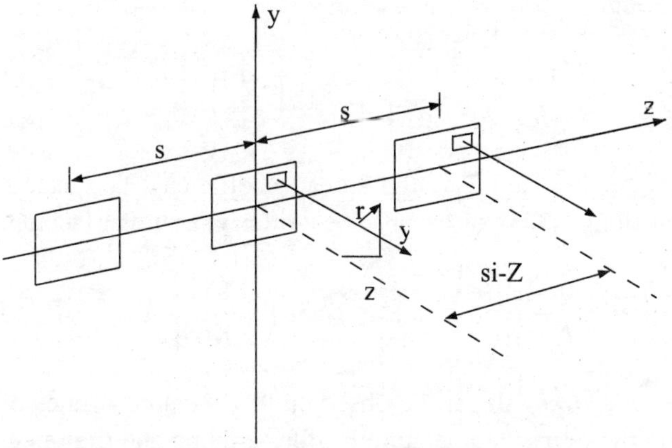

Figure 7.1 Definition sketch of an array of rectangular outlets

depend on the spacing of the outlet. In an array of circular outlets having a spacing $s = 3d$, the flow behaved like a plane jet when $x >\approx 12d$.

Superposition

N number of jets are symmetrically located about the z-axis, therefore, half of the number of jets, $j = \frac{N}{2}$, if N is even. For odd values of N, $j = \frac{(N-1)}{2}$ and the reference axis, z, passes though a jet located at the centre of the array. By adding the u^2 contribution of each individual jet, an equation can be written as

$$\frac{u_m^2}{U_0^2} = \frac{0.44A}{b_y b_z} \sum_{i=-j}^{i=+j} \exp\left\{-1.385\left(\eta_y^2 + \left(\frac{si}{b_z}\right)^2\right)\right\} \tag{7.11}$$

In Eq 7.11, i should include the value zero as well. It gives the distribution of the momentum flux for various y locations in the $z = 0$ plane. If we set $y = 0$ in Eq 7.11, the maximum of u_m^2, which is denoted as u_{m0}^2, will be given by the following:

$$\frac{u_{m0}^2}{U_0^2} = \frac{0.44A}{b_y b_z} \sum_{i=-j}^{i=+j} \exp\left\{-1.385\left(\frac{si}{b_z}\right)^2\right\} \tag{7.12}$$

Further, based on the error functions which were used in Eq 7.6, a modified expression in lieu of Eq 7.12 can be written as

$$\frac{u_{m0}^2}{U_0^2} = \left\{erf\left(\frac{B_0}{\sqrt{2}b}\right) erf\left(\frac{L_0}{\sqrt{2}b}\right)\right\} \sum_{i=-j}^{i=+j} \exp\left\{-1.385\left(\frac{si}{b_z}\right)^2\right\} \tag{7.13}$$

For an odd number of jets, the decay of the tracer flux along the x-axis will be similar to Eq 7.12, and

$$\frac{u_{m0}C_{m0}}{U_0 C_0} = \frac{0.44A}{b_{yc} b_{zc}} \sum_{i=-j}^{i=+j} \exp\left\{-1.385\left(\frac{si}{b_{zc}}\right)^2\right\} \tag{7.14}$$

where $b_{yc} = \lambda b_y$, $b_{zc} = \lambda b_z$ and λ is the spread coefficient. The excess temperature variation along the x-axis can also be written in a similar fashion,

$$\frac{u_{m0}\Delta T_{mo}}{U_0 \Delta T_0} = \frac{0.44A}{b_{yt} b_{zt}} \sum_{i=-j}^{i=+j} \exp\left\{-1.385\left(\frac{si}{b_{zt}}\right)^2\right\} \tag{7.15}$$

where $b_{yt} = \lambda b_y$ and $b_{zt} = \lambda b_z$. It can be observed that the spread rates of the concentration and the excess temperature profiles are one and the same. For even number of jets in the array, a minor modification will be necessary to account for the distance of each jet from the x–y plane of symmetry. Nonetheless, knowing $\frac{u_{m0}}{U_0}$, either from Eq 7.6 or Eq 7.10, the variation of $\frac{C_{m0}}{C_0}$ and $\frac{\Delta T_{m0}}{\Delta T_0}$ can be predicted based on Eqs 7.14 and 7.15, respectively.

7.2.3 Experimental Results: Stagnant Ambient

For circular jets, the spread coefficient c seems to depend on the conditions prevailing upstream of the nozzle in the plenum chamber. Therefore, the experimental values of different researchers do differ by some margin. An average value of $c \approx 0.098$ (in the relationship $b_{0.5} = cx$) was measured by Dash (1981). The location of the virtual origin was observed to be at 1.0d downstream of the efflux section. The value of c was necessary for predicting the theoretical curves pertaining to multiple jets tested in his set up. The theoretical variation of $\left(\frac{U_0}{u_m}\right)^2$ with distance $\left(\frac{x}{d}\right)$ were found from Eq 7.12. Agreement between the computed and observed values for different $N(= 1, 2, 3, 5$ and $7)$, with a spacing of $s = 3d$, seems to be good, as shown in Fig 7.2.

As N became large, the flow behaved as though it emerged from a plane slot of effective width $2b_e$. For such a case, $\left(\frac{U_0}{u_{m0}}\right)^2$ varies linearly with the distance $\frac{x}{d}$. Results of Knystautas (1964) for large values of N (not shown here) also exhibit similar behaviour. However, Knystautas assigned a much

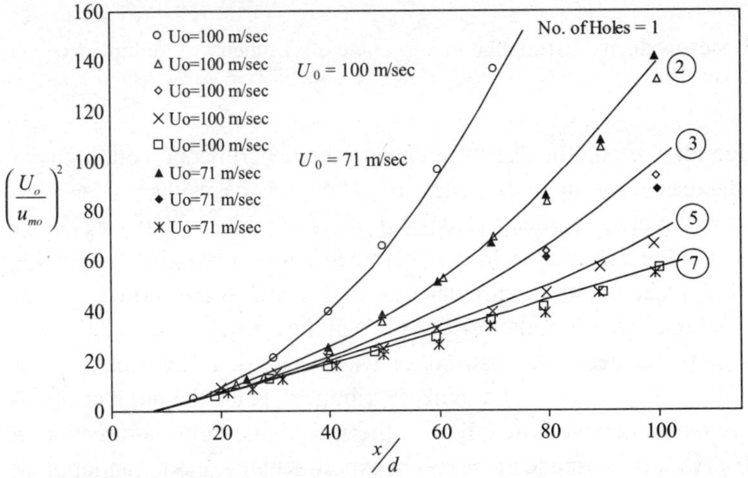

Figure 7.2 Variation of maximum dynamic pressure for multiple round jets (Source: Dash, 1981)

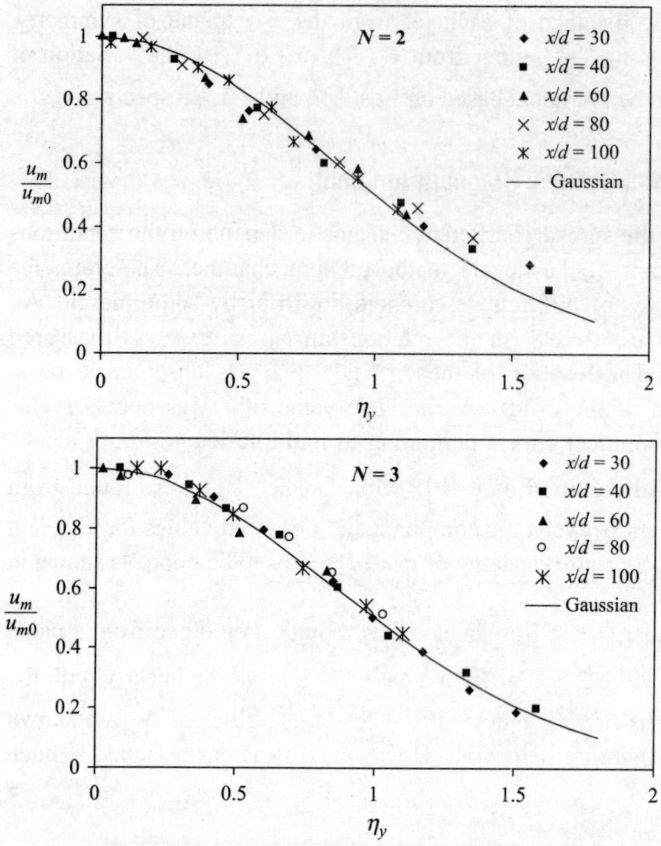

Figure 7.3 Mean velocity distribution in x–y plane of symmetry of multiple free jets (Source: Dash, 1981)

lower value for c ($= 0.09$). The difference in the experimental conditions is, perhaps, the reason for the discrepancy in the two sets of results.

In the $z = 0$ plane, the mean velocity profiles exhibit similarity as shown in Fig 7.3, and the distribution follows the theoretical curve given in Eq 7.9, for $\eta_z = 0$. Typical results for circular jets, $N = 2$ and 3, are shown, and the results are similar even for other values of N in the array.

Even though the theoretical distribution was meant for a single jet, it can be seen that multiple jets follow the same distribution. Thus, the number of jets in the array seems to have little effect on the spread rate in the y-direction. A similar observation was made in the recent experimental work of Yannopolous et al. (2006) on circular jets. Experiments on rectangular outlets of different aspect ratios also revealed a similar behaviour of the flow.

Velocity distribution in the *x–z* plane

The distribution of velocity in the x–z plane of symmetry, given in Eq 7.12, is shown in Fig 7.4. Results pertaining to three circular nozzles and three rectangular nozzles are shown in the diagram. Once again, the observed lateral distribution of the mean velocities, u and u_m, is seen to be in fair agreement with theory. The distribution of the mean velocity for other values of N, up to a maximum of 7, can be found elsewhere (Dash, 1981).

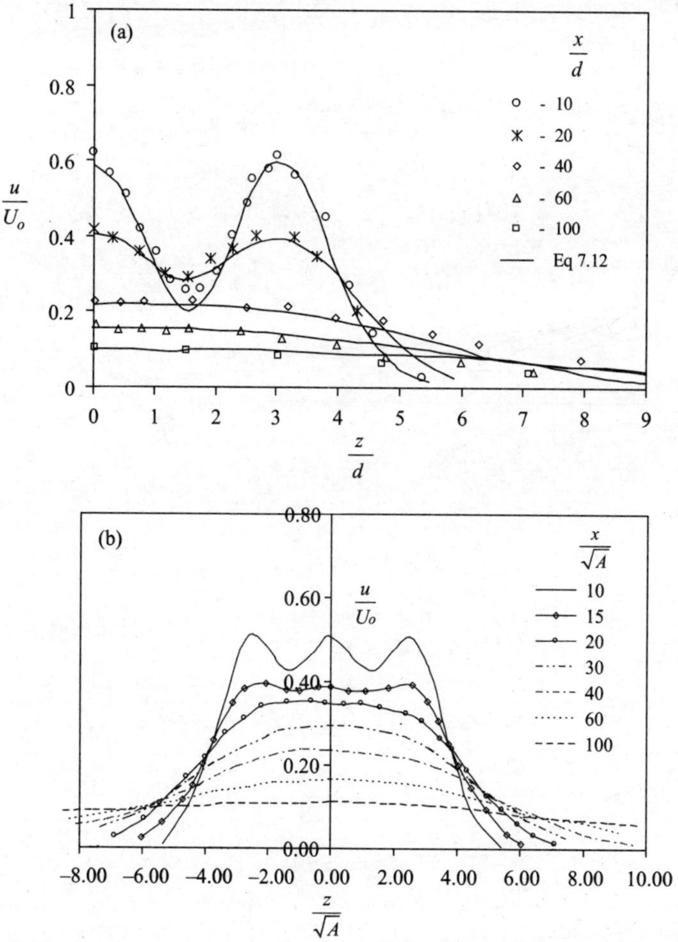

Figure 7.4 Mean velocity distribution in the x–z plane of multiple free jets (Source: Dash, 1981)

Decay of maximum velocity

The variation of the maximum velocity u_{m0} (which is the same as u_m in case of a round jet) along the x axis of an array of circular and rectangular outlets is shown in Fig 7.5, by considering $\sqrt{A}$ as the normalizing parameter for x. The curves look quite different beyond a short downstream location that marks the merger of the individual jets.

Soon after merger, the flow behaves as though it emerged from an equivalent area equal to N times A. Consequently, $\sqrt{NA}$ should be selected as a characteristic length scale for normalising the distance x. This is seen to be

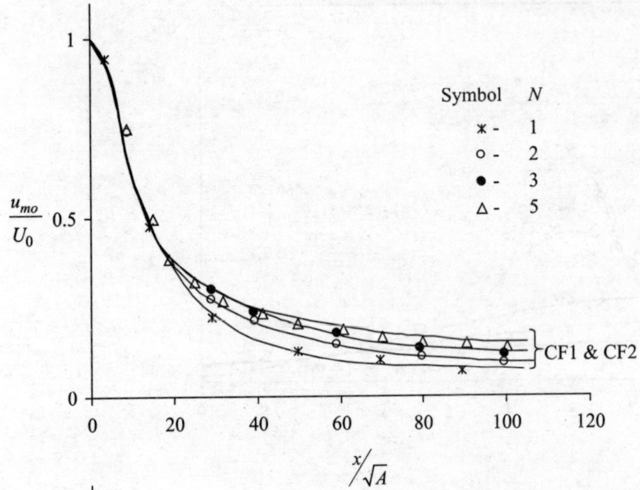

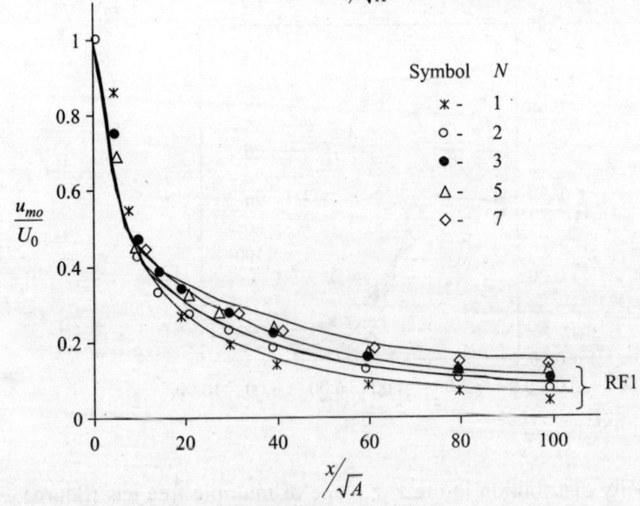

Figure 7.5 Maximum velocity decay for multiple jets (Source: Dash, 1981)

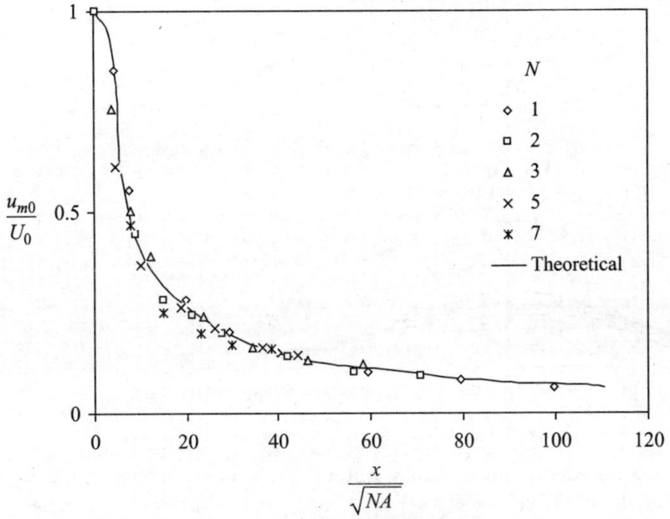

Figure 7.6 Normalised maximum velocity decay for multiple jets

true as depicted in Fig 7.6. A similar conclusion was arrived at based on a more rigorous analysis of the superposition method (Yannopolous et al., 2006).

Tracer distribution

Hodgson et al. (1999) measured the concentration field downstream of an array of 3 circular jets, spaced at $2d$ centre to centre. The jet Reynolds number was equal to 4900, and measurements were carried out for distances as far as $x \sim 85d$. While writing the expression for the decay of the maximum concentration of an individual jet, the value of λ was selected as 1.17. The relationships proposed for predicting the maximum velocity and the maximum concentration along the x-axis of a single jet are

$$\frac{u_m}{U_0} = \frac{6.13}{\left(\frac{x}{d}\right)} \tag{7.16}$$

and

$$\frac{C_m}{C_0} = \frac{5.34}{\left(\frac{x}{d}\right)} \tag{7.17}$$

Based on the superposition method, the distribution of the concentration in the x–z plane was predicted. The predicted distribution agreed reasonably well with the measured values in the developed flow region. As a typical illustration, Fig 7.7 demonstrates the results.

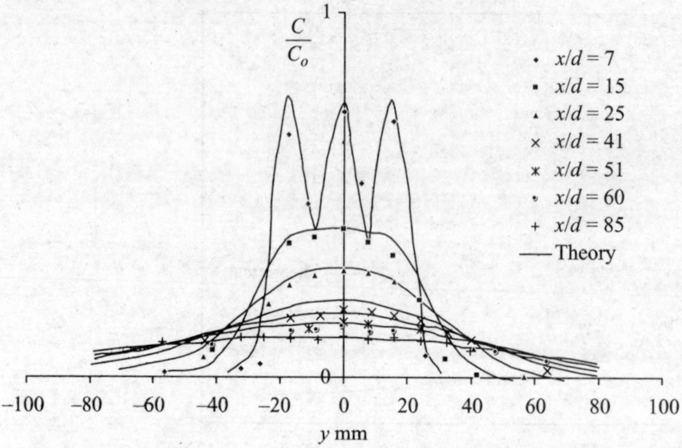

Figure 7.7 Concentration distribution across multiple jets in a stagnant ambient (Source: Hodgson et al., 1999)

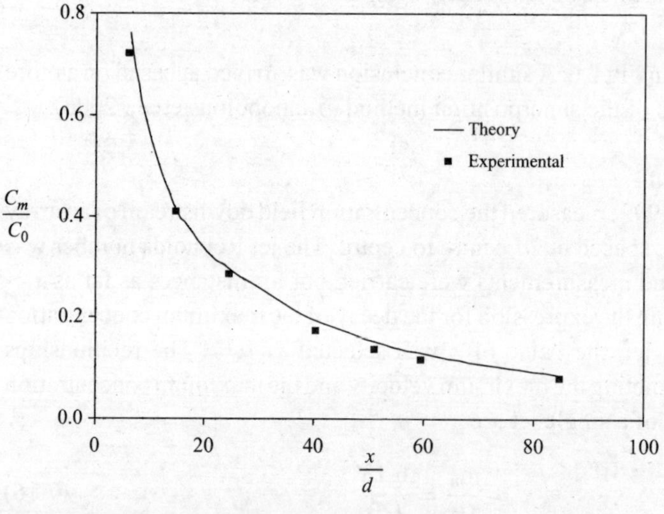

Figure 7.8 Decay of maximum concentration for multiple jets stagnant ambient (Source: Hodgson et al., 1999)

Decay of maximum concentration

The variation of the maximum mean concentration C_m is shown in Fig 7.8, and, here again, the agreement between observed and predicted values seems to be good. The desired accuracy of the concentration observed might have been compromised due to the technique followed in the experiments, that is, collecting samples from different locations before measuring their concentration.

Recent data for validation of superposition

Experiments were conducted with interacting round jets to validate the method of superposition (Yannopolous et al., 2006). The densimetric Froude number was in the range 1.8 to ∞ and the spacings adopted were 1.5 and 3 times the nozzle diameter. The centreline velocities were determined using a motion picture technique and the concentration was recorded by conductivity probes. By observing the intersect of optical boundaries, the merger effects of adjacent jets was noticed to begin at $x \simeq 2.5s$. However, the actual merger effect was noticeable after a distance of $x \simeq 5s$.

To avoid subjectivity in the definition of these distances, a well-defined criterion was proposed to locate them (Pani, 2008). Merger of jets begins at a distance x, where nearby jet/jets contribute 1 per cent of the momentum of a single jet. For $N = 2$ and 3, this merger distance was predicted as $5.5s$ and $5.1s$, respectively. Similarly, by equating the momentum flux of a slot jet spanning over two adjacent openings with that of a single circular jet, the distance at which the flow from the array of nozzles behaves like a slot jet was predicted as $6.6s$, approximately. The difference between the two distances ($\approx 1.5s$) represents the zone in which the process of merger gets completed.

The flow field of vertically directed multiple jets, held in a line, can have 5 zones based on depth of submergence and number of jets, N. The 5 zones, in sequence, are: initial symmetric flow of individual jets, lower transition zone, a two-dimensional decay zone, upper transition zone and the reversal to an axially symmetric flow of the combination of jets.

Based on dimensional arguments, a proximity number $P_n = \left(\frac{s}{2d}\right) F_0^{-1}$ has been introduced as an additional parameter which affects the behaviour of merging jets. For $N > 2$, the proximity number acts as a similarity parameter. Hence, it will suffice if the P_n in the model and prototype of multiport diffusers is made the same. Therefore, the number of ports provided in the model could be much smaller than the number in the prototype.

7.3 Multiple Jets in a Coflowing Stream

The momentum flux resulting from an array of jets in a coflowing stream can be predicted at any location by superposing (i.e., adding) quantities like $\frac{u_m}{U_0}\frac{\Delta u_m}{\Delta U_0}$ and $\frac{u_m}{U_0}\frac{C_m}{C_0}$ of the individual jets. For example, the resulting excess momentum flux for N jets can be written as

$$\left(\frac{u_m \Delta u_m}{U_0 \Delta U_0}\right)_N = \sum_N \frac{u \Delta u}{U_0 \Delta U_0} \tag{7.18}$$

Normally, it would be desirable to predict $\left(\frac{\Delta u_m}{\Delta U_0}\right)_N$ or $\left(\frac{\Delta u_m}{U_a}\right)_N$ as a function of the distance $\frac{x}{\theta}$. By following the procedure akin to the case of a single jet, in case of multiple jets, the expression can be written as

$$\left(\frac{u_m}{U_0}\frac{\Delta u_m}{\Delta U_0}\right)_N = \beta q_N + (1-\beta) q_N^2 \tag{7.19}$$

The velocity ratio, $\beta = \frac{U_a}{U_0}$, is independent of N. Importantly, in Eq 7.19, q_N represents the ratio $\left(\frac{\Delta u_m}{\Delta U_0}\right)_N$ for multiple jets. Substituting for the quantity on the left hand side of Eq 7.19, from Eq 7.18, the quadratic equation can be solved for q_N.

Note that in solving this problem, we are evaluating a single unknown, q_N, with the presumption that the spread coefficient of an individual jet, $\sqrt{2}c$, is known at the x-location. The accuracy of results will, therefore, depend on how good a value is assigned to $\sqrt{2}c$ in the computations (Section 6.6.1). The coefficient $\sqrt{2}c$ is a function of the velocity ratio, β and local $\frac{\Delta u_m}{U_a}$. The value of q_N is needed to evaluate $\left(\frac{\Delta u_m}{U_a}\right)_N$. Therefore, the method to evaluate $\left(\frac{\Delta u_m}{U_a}\right)_N$ is also included. A step-by-step computational procedure is suggested below.

7.3.1 (A) Prediction of $\left(\frac{\Delta u_m}{U_a}\right)_N$

1. An appropriate value for $\sqrt{2}c$ is selected/computed for the given β.
2. $\frac{u_m}{U_0}\frac{\Delta u_m}{\Delta U_0}$ for a single circular jet is found from Eq 6.45 (counterpart of Eq 7.10 in stagnant ambient) at the specified distance x. If rectangular/square nozzles are involved, Eq 6.50 (counterpart of Eq 7.6 in stagnant ambient) will be used.
3. For each of the jets in the array, one can find $\frac{u}{U_0}\frac{\Delta u}{\Delta U_0} = \frac{u_m}{U_0}\frac{\Delta u_m}{\Delta U_0} \exp\left\{-\left(\frac{r}{\sqrt{2}cx}\right)^2\right\}$. The radial distance $r (= z)$ is the distance of the ith jet axis from the mid-plane (i.e. $z = 0$).
4. The quantities of step 3 are added to get $\left(\frac{u_m}{U_0}\frac{\Delta u_m}{\Delta U_0}\right)_N$. The outlets close to the axis of symmetry contribute the most to the summation; the role of the far placed ones will be nearly zero.
5. By setting $a_0 = -\left(\frac{u_m}{U_0}\frac{\Delta u_m}{\Delta U_0}\right)_N$; $a_1 = \beta$; and $a_2 = (1-\beta)$ in Eq 7.19, we have

$$a_0 + a_1 q_N + a_2 q_N^2 = 0 \tag{7.20}$$

6. The quadratic Eq 7.20 is solved for q_N.
7. Therefore, $\left(\frac{u_m}{U_0}\right)_N = -\frac{a_0}{q_N}$.
8. Having found $\left(\frac{u_m}{U_0}\right)_N$, compute $\left(\frac{\Delta u_m}{U_a}\right)_N = \frac{1}{\beta}\left(\frac{u_m}{U_0}\right)_N - 1$.
9. For a new location x, steps 1 through 8 are repeated.

Recall that the distribution of the tracer flux is $\frac{uC}{u_m C_m} = \exp\left(-\frac{\zeta^2}{2}\right)$, and the maximum value of C decays in accordance with

$$\frac{u_m}{U_0}\frac{C_m}{C_0} = 1 - \exp\left\{-\frac{1}{4}\left(\frac{d}{\lambda\sqrt{2c}x}\right)^2\right\} \tag{7.21}$$

The steps involved for prediction of the maximum concentration decay with distance x have been summarised below.

7.3.2 (B) Prediction of concentration C_m

1. For known values of λ, $\sqrt{2}c$, d and x, the value of $\frac{u_m}{U_0}\frac{C_m}{C_0}$ for a single jet is found from Eq 7.21.
2. Therefore, $\frac{u}{U_0}\frac{C}{C_0} = \frac{u_m}{U_0}\frac{C_m}{C_0}\exp\left(-\frac{\zeta^2}{2}\right)$.
3. The contribution of each outlet is added, resulting in $\left(\frac{u_m C_m}{U_0 C_0}\right)_N = \sum_N \frac{uC}{U_0 C_0}$.
4. Since $\left(\frac{u_m}{U_0}\right)_N$ is known from step 7 of (A), the concentration decay in case of 'N' jets is predicted as $\left(\frac{C_m}{C_0}\right)_N = \frac{\left(\frac{u_m C_m}{U_0 C_0}\right)_N}{\left(\frac{u_m}{U_0}\right)_N}$.
5. Steps 1 through 4 are repeated for a new x after completing Part (A).

In the above computations, the underlying assumption is that the contribution of the fluctuating components is negligible. Further, it is evident that the theoretical values for multiple jets in a coflowing stream depend upon the number of jets in the array, the port spacing and the velocity ratio β. The explicit role of β on the centreline dilution S_c can be kept out of picture if it is expressed in the following form,

$$\frac{\beta}{1-\beta}S_c = \frac{\lambda^2}{\left(\frac{\Delta u_m}{U_a}\right)_N} \frac{\sum_{-n}^{n}\left\{e^{-\frac{1}{2}\left(\frac{is_p}{b}\right)^2}\right\}}{\sum_{-n}^{n}\left\{e^{-\frac{1}{2}\left(\frac{is_p}{\lambda b}\right)^2}\right\}} \tag{7.22}$$

where s_p is the port spacing.

7.3.3 Effects of Boundary

So far, we have superposed the excess momentum flux and the flux of the tracer from N jets, with an implicit assumption that the expanse of the ambient fluid is very large. In practice, jets discharge into water bodies of limited depth/width. Assuming that the channel is fairly wide, the effects of the free surface and the channel bottom can be estimated. The procedure relies on the principle of superposition treating the free surface and the channel bottom as ideal reflectors. For illustration, let H be the water body depth and the array of the jets are held at a distance of $\frac{H}{2}$ above the bottom. Using the method of images, the maximum excess velocity and the dilutions were predicted.

Depending on the relative depth $\frac{H}{d}$, the presence of the boundary will affect the excess momentum flux and the flux of the tracer. The resulting dilution, far downstream of the multiple jets, will level off at significantly lower values than indicated by the ideal case. The ideal case is the one where effects of boundary are not taken into consideration. There are no systematic observations on this aspect.

7.3.4 Validation of Theory with Experiments

Variation of the excess velocity and the concentration of multiple jets, $N = 11$ and 6, were measured for a range of β values (Wang, 1996 and Wang, 2000). The variation of the excess velocity along the central axis of the arrays is shown in Fig 7.9. The excess velocity observed experimentally agrees with the predicted values based on the superposition of excess momentum quite well. In contrast, the predictions based on the addition of the excess velocity, the dotted

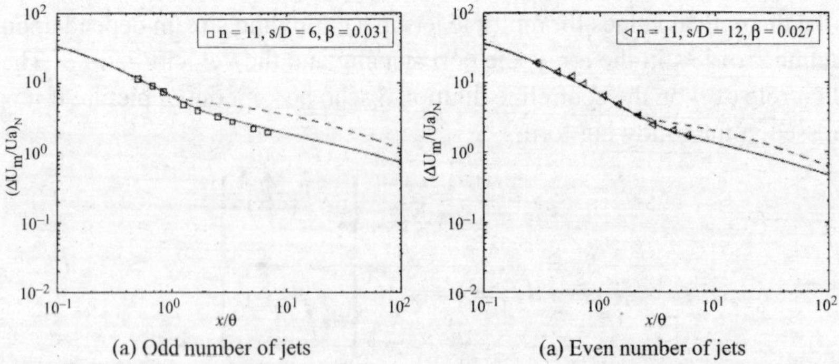

Figure 7.9 Velocity decay of multiple coflowing jets (Data from Wang, 2000)

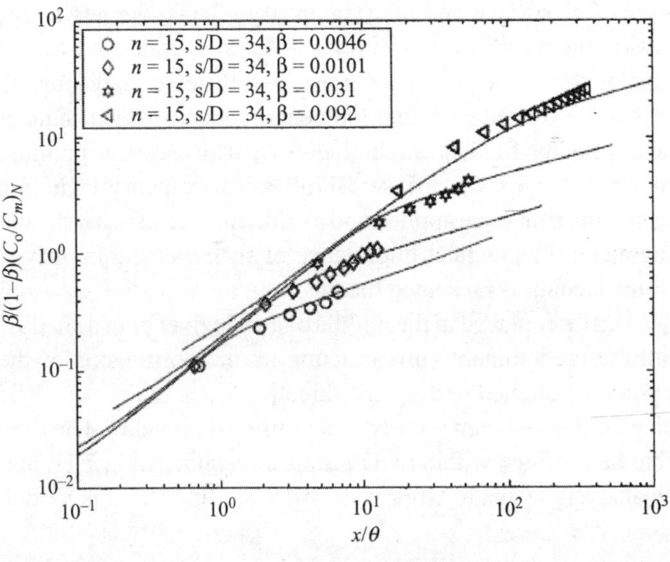

Figure 7.10 Streamwise variation of dilution for multiple coflowing jets (Data from Wang, 2000)

line in the figure, are not satisfactory. This lends support to the efficacy of the modified Reichardt's hypothesis (Section 6.6) for multiple coflowing jets.

Observations on the concentration variation of an array of 15 jets are available for four velocity ratios lying in the range 0.0046–0.092. These are shown in Fig 7.10, along with the theoretical curves based on the method of superposition. The agreement between theory and the observed values seems to be remarkable.

7.4 Momentum Driven Multiple Jets in Shallow Water

Heated water from large steam electric generating stations needs to be diluted as quickly as possible by mixing it with an ambient fluid due to economic reasons. The flows involved are usually large and multiport diffusers seem to be a reasonable choice. The jet velocities near the port could be as high as 5 m/s, and the flow is momentum driven. The type of diffuser to be adopted for each power plant is site specific and the arrangement of the ports in the diffuser could be different. The length of a diffuser depends on many factors including permissible temperature rise of the ambient, station heat rejection rate, discharge velocity, ambient current, water depth etc.

In the absence of free surface and other boundary effects, the jets merge and the flow outside the shear layer is similar to the one generated by the axially symmetric or the plane jet. In contrast, a diffuser of large length in shallow water can be treated as a line momentum source which induces a two-dimensional velocity field in the ambient. In such cases, a pressure discontinuity is generated across the diffuser. The classical jet theory, including the point-source method, cannot be applied and a different type of analysis will be needed. Recognizing the practical importance of such momentum driven multiple jets, a brief account is presented below.

Let us consider a diffuser placed in the middle of a wide river or in a shallow coastal zone with the predominant current acting normal to the axis of the diffuser. All the ports are aligned in the same direction along the current. The entire system behaves like a jet pump or a two-dimensional propeller. The flow can be imagined to be confined within two limiting streamlines, Fig 7.11, and the inviscid flow analysis seems to work well (Adams, 1982). In this model,

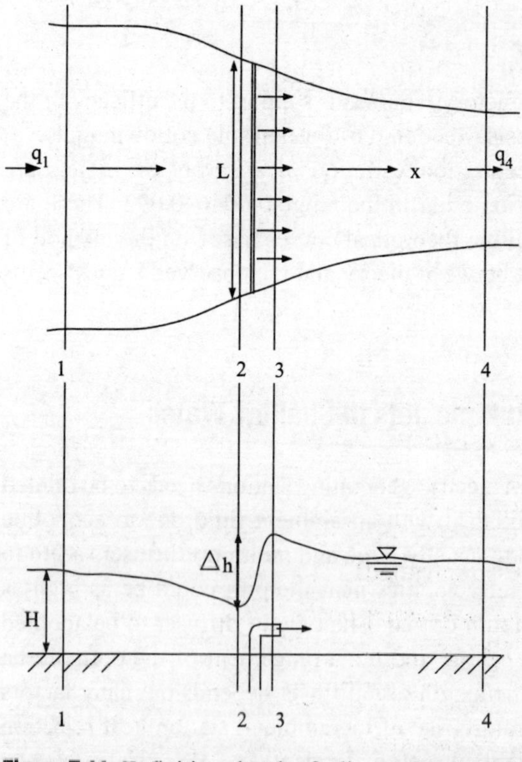

Figure 7.11 Definition sketch of a line momentum source

the flow and the water depth outside the limiting streamlines are not effected by the flow induced by the diffuser.

A simple application of the inviscid flow helps to write Bernoulli's equation along streamlines behind and in front of the diffuser. The differential head imparted by the diffuser, Δh, was found from the momentum equation. Based on the assumption that $H >> \Delta h$ and the induced flow Q is very large compared to the diffuser discharge, Q_0, an expression for the dilution was derived. The condenser water dilution, S, can be predicted from the following:

$$S = \frac{U_a A}{2Q_0} + \sqrt{\left(\frac{U_a A}{2Q_0}\right)^2 + \frac{A\cos\alpha}{2Na_0}}, \tag{7.23}$$

where $A = HL$; N = number of ports; a_0 = area of port; U_a = current speed upstream of the diffuser and α = angle which the ports make with the horizontal.

Theoretically, the coefficient of contraction of the limiting streamlines is equal to $\left(0.5 + 0.5\frac{U_a}{U_4}\right)$, where U_4 is the velocity at the section having nearly parallel streamlines. Thus, for $U_a = 0$, the coefficient of contraction = 0.5, and the diffuser dilution is reduced by a factor of 2. The measured and the predicted dilutions, Eq 7.23, seem to agree well as shown in Fig 7.12. The

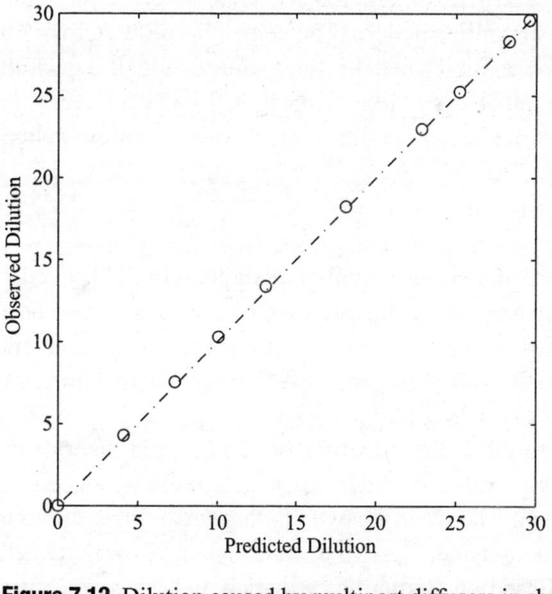

Figure 7.12 Dilution caused by multiport diffusers in shallow water

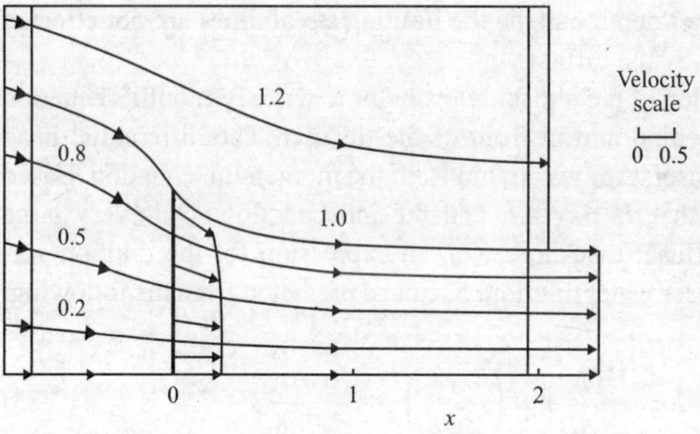

Figure 7.13 Streamline pattern for multiport diffusers in shallow water ($U_a = 0.286$ m/s).

scatter of the experimental data points being very small is not shown in the figure.

A more rigorous treatment of the flow based on slip streamlines and distribution of vortices along them (both unknown quantities) helped to predict the pattern of streamlines for a range of flow variables (Lee and Greenberg, 1984). A typical plot of the streamlines obtained for $U_a = 0.286$ m/s is shown in Fig 7.13. The slip streamlines ($\psi = 1$) and the resulting flow contraction that occurs has been highlighted in the figure. Experimental observations agree with the predictions of the streamline pattern. The velocity profiles across the converged section of the flow exhibit a near uniform value over half the width of the line source and then fall sharply at the two edges. This occurs because of the real fluid flow effects not accounted for in the theoretical formulation.

7.4.1 Shoreline Effects

Figure 7.14 shows the layout of a diffuser parallel to the shoreline. The feeder pipe is at right angles to the axis of the diffuser. Such a layout is known as a T-diffuser. The flow features are also depicted in the figure. As the ambient water available from behind the diffuser is restricted, proximity of the shoreline reduces the dilution that can be achieved.

In an approach similar to the coflowing diffuser, the inviscid theory was applied to predict the resulting dilution (Adams, 1982). The flow induced by the diffuser was assumed to pass between the two limiting streamlines, without having any influence on the outside. Here, the water depth, H, was treated as shallow and constant. Based on Bernoulli's equation and the momentum

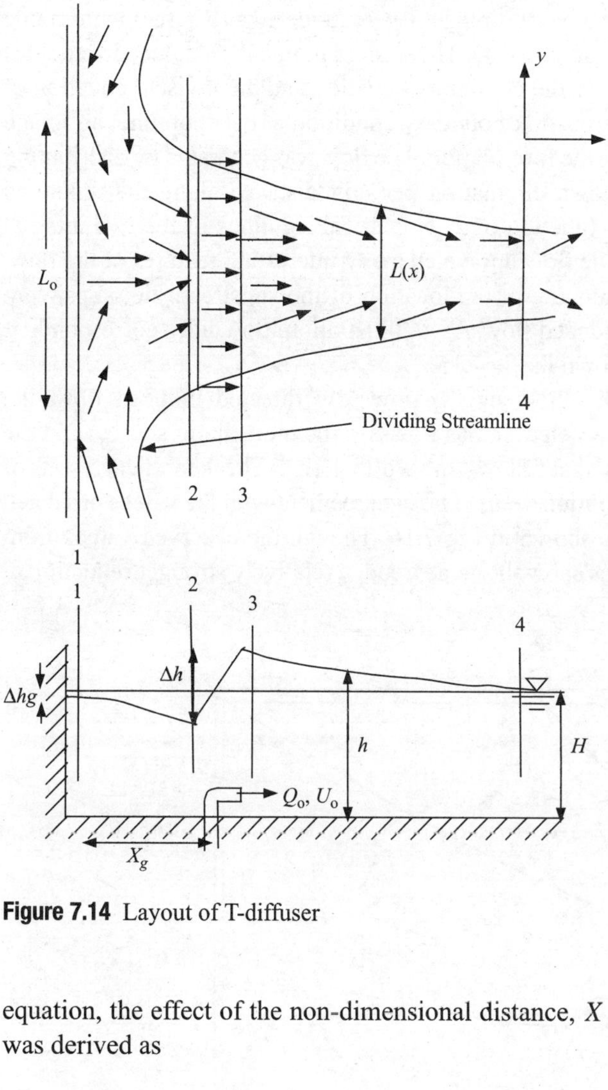

Figure 7.14 Layout of T-diffuser

equation, the effect of the non-dimensional distance, $X = \frac{x_s}{L}$, on the dilution was derived as

$$\left(\frac{S}{S_\infty}\right)^2 = 2\left\{\left(\frac{X}{c}\right)^2 + \left(\frac{X}{c}\right) - \frac{X}{c}\sqrt{\left(\frac{X}{c}\right)^2 + 2\left(\frac{X}{c}\right)}\right\} \tag{7.24}$$

where c (< 1) is an empirical constant and S_∞ is the dilution when the shoreline is far removed from the location of the T-diffuser. By trial, it was found that a value of $c = 0.15$ brings close agreement between the observed and the predicted dilutions from Eq 7.24.

In a different approach, multiple jets were modeled as a x-momentum line source (Lee and Greenberg, 1984). The desired momentum induced velocities u and v are governed by the continuity equation and the inviscid equation of motion. To satisfy the no-flux boundary condition at the shoreline, an image needs to be included in the formulation. The flow was generated by distributing line vortices of unknown circulation per unit distance along the unknown slipstream which is a function of x. Numerical solutions of the two integral equations governing the flow unraveled many interesting features of the flow, such as the streamline pattern, the location of the slipstream lip, slipstream contraction and the induced flow. As an illustration, flow lines corresponding to $\frac{x_s}{L} = 0.0828$ are shown in Fig 7.15.

The features of the flow, like an inwardly directed induced flow, the sharp turns, the flow acceleration as it passes the momentum source, and the significant contraction that follows, are worth noting. The theoretical results of this formulation on dilution seem to be in agreement with the values predicted from Eq 7.24. This is shown in Fig 7.16. Though the observed values from different investigations show the same trend, a relatively stronger reduction in

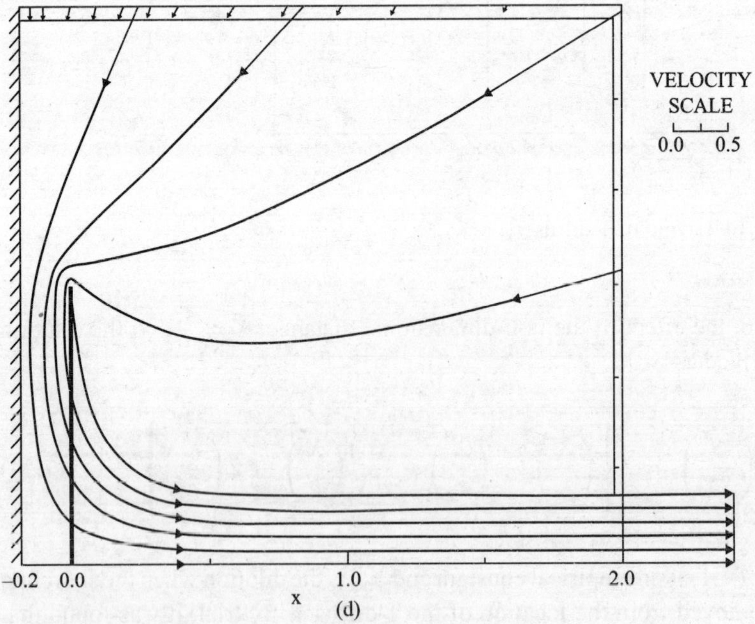

Figure 7.15 Flow pattern for multiport diffusers: Line vortices method (Source: Lee and Greenberg, 1984)

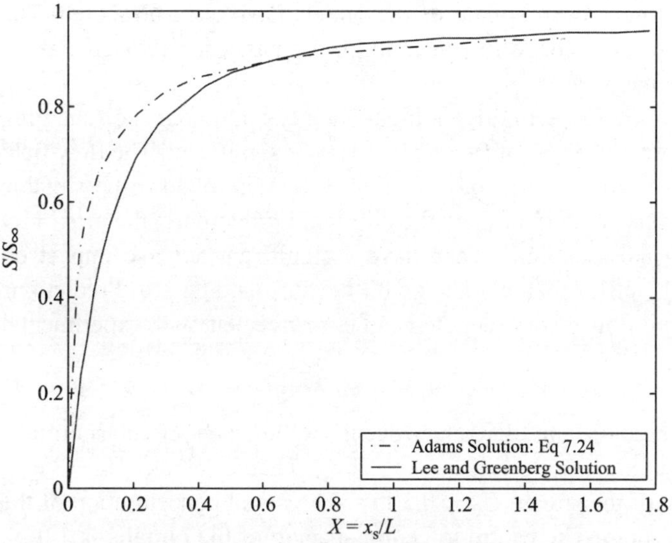

Figure 7.16 Dependence of dilution on the distance x_s

the induced flow is indicated. The difference between theory and the observed values is of the order of 10 per cent.

The theory can be useful if the following conditions are satisfied:

(a) $\frac{L}{H} >> 1$, port spacing $\leq H$
(b) $\frac{L}{H} \leq 100$, so that boundary shear can be neglected
(c) the induced flow is large compared to the source flow

7.5 Summary

The distribution of the momentum flux downstream of an array of jets in a stagnant environment can be predicted based on the method of superposition. As a result, the decay of quantities like the maximum velocity, the maximum tracer concentration or the maximum excess temperature can be predicted along the axis of symmetry of the array. Experimental results in a stagnant environment of large expanse amply validate the predictions from the method of superposition.

For multiple jets in a coflowing environment, the basic method of superposition helps to predict the decay of the maximum excess velocity and the maximum concentration. It is an extension of the single coflowing jet to account for multiple sources of flow. Simple methods to predict the excess velocity and dilution down stream of the outlets have been presented. The method of superposition for coflowing multiple jets has been validated making use of experimental data.

In the event of momentum driven, high-velocity multiple jets discharging into shallow water, the effect of free surface is very significant. The flow field behaves like a jet pump. The potential flow theory is found to be suitable for prediction of the streamlines and the dilution. Furthermore, the proximity of a boundary, like a shoreline, can have a significant adverse impact on the mixing of the effluent (hot water) with the surrounding sea. Predictions based on potential flow theory seem to be in fair agreement with experimental observations.

Example 7.1: Three 0.3 m diameter, deeply submerged circular outfalls discharge effluent into a coflowing stream. The discharge velocity $U_0 = 3$ m/s, and the velocity of the stream $U_a = 0.5$ m/s. The initial concentration of the tracer $C_0 = 500$ ppm. The centre-to-centre spacing of the outfalls is 1.0 m. Find, at intervals of 5 m, the approximate variation of the jet thickness, the excess velocity, and the average dilution for the first 25 m.

Solution: The initial flow conditions are identical to the single port discharge of example 6.4. We shall, therefore, borrow the results from the said example in the process of predicting the characteristics of the multiple jets. As an illustration, a set of computations for $x = 10$ m is shown in tabular form. The final values of b_g, Δu_m and S_{av} are also shown in a separate table.

Given, $U_a = 0.5$ m/s, $U_0 = 3.0$ m/s, $d = 0.3$ m, port spacing $s = 1.0$ m. Assume $\lambda = 1.2$.

Hence, $\Delta U_0 = 2.5$ m/s, and $A_0 = 0.0707$ m^2. The excess momentum $M_{0e} = 0.0707 \times 3.0 \times 2.5 = 0.530$ m^4/s^2.

$$\theta = \sqrt{\frac{M_{e0}}{U_a^2}} = \sqrt{\frac{0.530}{0.5^2}} = 1.456 \text{ m}.$$

For $x = 10$ m, $\frac{x}{\theta} = 6.868$ and the corresponding $\frac{b_{gc}}{\theta} = 0.5123$.

Hence, $b_g = \frac{b_{gc}}{1.2} = \frac{0.5123 \times \theta}{1.2} = \frac{0.5123 \times 1.456}{1.2} = 0.6216$ m.

For a single jet, Example 6.4, at $x = 10$ m, $\frac{\Delta u_m}{U_a} = 1.1193$.

For multiple jets, we should only superpose the values of $u \Delta u$, i.e. $(U_a + \Delta u)\, \Delta u$, but not the excess velocity Δu as was sometimes done in the past.

Hence we can write,

$$\left(\frac{u_m \Delta u_m}{U_a^2}\right)_3 = \sum_{-1}^{+1} \left(\frac{\Delta u_m}{U_a}\right)_{\text{single}} \mathrm{e}^{-\left(\frac{is}{b_g}\right)^2} \times \left\{1 + \left(\frac{\Delta u_m}{U_a}\right)_{\text{single}} \mathrm{e}^{-\left(\frac{is}{b_g}\right)^2}\right\} = SUM3. \quad (7.25)$$

Location $x = 10$ m:

i	$\frac{\Delta u_m}{U_a}$	$\left(\frac{\Delta u_m}{U_a}\right)\mathrm{e}^{-\left(\frac{is}{b_g}\right)^2}$	$\left\{1+\left(\frac{\Delta u_m}{U_a}\right)\mathrm{e}^{-\left(\frac{is}{b_g}\right)^2}\right\}$	$(Col.3 \times Col.4)$
0	1.1193	1.1193	2.1193	2.3721
−1	1.1193	0.0841	1.0841	0.09117
+1	1.1193	0.0841	1.0841	0.09117
				SUM3 = 2.55444

We can also rewrite,

$$(u_m \Delta u_m)_3 = \left\{\left(\frac{\Delta u_m}{U_a}\right)_3^2 + \left(\frac{\Delta u_m}{U_a}\right)_3\right\} U_a^2 \quad (7.26)$$

From Eqs 7.25 and 7.26, $\left(\frac{\Delta u_m}{U_a}\right)_3^2 + \left(\frac{\Delta u_m}{U_a}\right)_3 = SUM3$. The positive root of the equation is, $\left(\frac{\Delta u_m}{U_a}\right)_3 = \frac{-1+\sqrt{1+4a_0}}{2}$, where $a_0 = SUM3$.

Hence, $\left(\frac{\Delta u_m}{U_a}\right)_3 = \frac{-1+\sqrt{1+4\times 2.5544}}{2} = 1.1746$.

Had we superposed the velocity excess, the sum of the values in Column-3 will yield the required answer as $\left(\frac{\Delta u_m}{U_a}\right)_3 = 1.2875$. The error is of the order of +9.6 per cent. Further downstream, the errors will be even more.

The final results at various locations along the axis of the jet is shown in the following table. The values of $\frac{\Delta u_m}{U_a}$ based on the superposition of the velocities are also shown in the last column.

x (m)	$\frac{x}{\theta}$	b_g, m	$\frac{\Delta u_m}{U_a}$	$\left(\frac{\Delta u_m}{U_a}\right)_{vel.sup}$
5.0	0.434	0.3788	2.226	2.229
10.0	6.868	0.6216	1.174	1.287
15.0	10.302	0.7923	0.910	1.089
20.0	13.736	0.9341	0.785	0.974
25.0	17.170	1.0903	0.683	0.859
30.0	20.604	1.3040	0.573	0.716

For evaluating the centre-line concentration/dilution on the axis of symmetry of the array of three jets, the following expression based on the method of superposition can be used.

$$\frac{C_m}{C_0} = \frac{\beta}{1-\beta} \frac{\left(\frac{\Delta u_m}{U_a}\right)_N}{\lambda^2} \frac{\sum_{-n}^{n}\left\{e^{-\frac{1}{2}\left(\frac{is}{\lambda b}\right)^2}\right\}}{\sum_{-n}^{n}\left\{e^{-\frac{1}{2}\left(\frac{is}{b}\right)^2}\right\}} \tag{7.27}$$

In the above equation, i takes the values of 0, −1 and +1. The value $\left(\frac{\Delta u_m}{U_a}\right)_N$ corresponds to the value computed for the three jets, and is available from the earlier computations, shown in the above table. In Eq 7.27, the spacing $s = 1.0$ m, and $C_0 = 500$ ppm. The computed concentration and dilution at various x-distances are as follows.

x, m	$\frac{C_m}{C_0}$	C_m, ppm	S_c, Dilution
5	0.343	171.65	2.91
10	0.190	94.80	5.27
15	0.151	75.38	6.63
20	0.129	64.57	7.74
25	0.110	55.13	9.07
30	0.090	44.91	11.13

Example 7.2: In a stagnant ambient the velocity distribution is often described by the relationship $\frac{u}{u_m} = e^{-\left(\frac{r}{b_g}\right)^2}$ and the typical value for $b_g = 0.109 \times x$. An alternate distribution was derived based on Reichardt's hypothesis and is given as $\frac{u^2}{u_m^2} = e^{-\frac{1}{2}\left(\frac{r}{b}\right)^2}$. In order that either method will lead to the same distribution in agreement with the observed data, one can write a simple relationship

$b = 0.5 \times b_g = 0.054 \times x$. However, for the case of the coflowing stream, no such simple transformation exists between the two length scales. Based on the principle that the excess momentum of a jet is conserved, derive a general relationship between b and b_g, valid in the strong as well as the weak-jet region.

Solution: The excess momentum flux over the cross section of a round jet can be written as,

$$M_e = \int_0^\infty 2\pi r dr u \Delta u. \tag{7.28}$$

Substituting for $\Delta u = \Delta u_m e^{-\left(\frac{r}{b_g}\right)^2}$ in Eq 7.28, on integration, we have

$$M_e = \frac{\pi}{2}\Delta u_m^2 b_g^2 + \pi U_a \Delta u_m b_g^2. \tag{7.29}$$

Again, we can re-evaluate Eq 7.28 by substituting for $u\Delta u$ from Reichardt's hypothesis. Thus,

$$M_e = \int_0^\infty 2\pi r dr u_m \Delta u_m e^{-\frac{1}{2}\left(\frac{r}{b}\right)^2} = 2\pi b^2 u_m \Delta u_m. \tag{7.30}$$

Equating the two equations, Eqs 7.29 and 7.30, on simplification we can arrive at the following:

$$\frac{b}{b_g} = \sqrt{\frac{\left(\frac{\Delta u_m}{U_a}\right) + 2}{4\left(\frac{\Delta u_m}{U_a} + 1\right)}} = k_l. \tag{7.31}$$

The multiplying factor, k_l, designated as a link factor, changes along the trajectory of the jet. This is so, since $\frac{\Delta u_m}{U_a}$ is getting smaller and smaller with the distance x. For a stagnant ambient, or even in case of a strong jet, $k_l \sim 0.5$. In contrast, for the weak-jet case, $k_l \sim 0.707$. In the transition zone, k_l assumes values which lie between the two said limiting values.

Thus, the spreading hypothesis applicable to b can be written as

$$\frac{db}{dx} = k_l \left(c_g\right)_0 \left\{ \frac{\frac{\Delta u_m}{U_a}}{2 + \frac{\Delta u_m}{U_a}} \right\} \tag{7.32}$$

In the above equation, $(c_g)_0$ is the value of c_g in a stagnant ambient. Normally, a value of 0.109 is assigned to $(c_g)_0$.

Example 7.3: In place of a single 0.3 m diameter outfall, it is proposed to provide an alternate design involving multiple ports. The flow velocities in both the cases are the same, that is, $U_0 = 3.0$ m/s, and $U_a = 0.5$ m/s. Assume the spread coefficient $\lambda = 1.2$ and the distributions of the excess velocity and the concentration are Gaussian in nature. Three circular nozzles, $d = 0.173$ m each having a centre-to-centre spacing of 0.6 m may be selected for assessing the characteristics of the flow field. Contrast the results with the 0.3 m diameter jet.

Solution: If hand computations are done, the answers can be found by treating the problem in three parts: the single jet solution, superposition of momentum excess to find $\frac{\Delta u_m}{U_a}$ of multiple jets and superposition of the tracer flux for evaluation of the centre-line concentration/dilution of the multiple jets. This is identical to the steps followed in Example 7.1.

Single jet: Given, $d = 0.173$ m. Hence, $A_0 = \frac{\pi}{4} \times d^2 = 0.02356\ \text{m}^2$, and $Q_0 = A_0 U_0 = 0.02356 \times 3.0 = 0.07068\ \text{m}^3/\text{s}$.

Momentum excess, $M_{e0} = Q_0 \Delta U_0 = 0.07068 \times 2.5 = 0.1767\ \text{m}^4/\text{s}^2$. Thus, $\theta = \sqrt{\frac{0.1767}{0.5^2}} = 0.8407$ m.

The diameter of the jets being small, we shall evaluate the required quantities at intervals of 2 m along the longitudinal axis of the jet. Hand computation being laborious, the algorithm presented in Section 7.3.1 is used to find the required quantities by running simple computer programs. The results are shown for an array of three circular jets.

Multiple jets:

x, m	$\frac{x}{\theta}$	b_g, m	$\frac{\Delta u_m}{U_a}$	*Dilution*, S_c	$\frac{C_m}{C_0}$	C_m, ppm
2	2.382	0.231	2.354	3.057	0.327	163.5
4	4.764	0.349	1.436	4.879	0.205	102.5
6	7.146	0.440	1.080	6.019	0.166	83.0
8	9.528	0.516	0.916	6.754	0.148	74.0
10	11.911	0.581	0.825	7.382	0.135	67.7
12	14.293	0.639	0.764	7.983	0.125	62.6
14	16.675	0.691	0.716	8.572	0.116	58.0
16	19.057	0.738	0.678	9.149	0.109	54.5
18	21.439	0.782	0.644	9.715	0.103	51.5
20	23.822	0.822	0.615	10.269	0.097	48.5

It may be noticed that in the above example, the maximum value of the concentration at any given downstream location is not very different when three outlets discharge the same volume of effluent into the coflowing stream as that of an equivalent single large jet (see Example 6.4). This occurs because of merging of the adjacent jets at a short distance from the efflux section, the spacing of the jets $\frac{s}{d}$ being small. However, if a larger spacing of, say $\frac{s}{d} = 10$, is selected, better mixing can be achieved with multiple jets. Calculations show that the centreline dilution at $x = 20$ m will be 18.22, resulting in a significant jump in the dilution shown in the above table.

8 Potential Core and Virtual Origin

8.1 General

Mixing and diffusion of various types of jets have been discussed in the earlier chapters. The emphasis was on the far field, the 'fully developed region' of the jet. Little attention was paid to the flow in the near field where the flow is 'developing'. In many applications, especially in environmental and hydraulic engineering, this seems to be justified as only the mixing in the far field is of consequence. Nonetheless, the near field plays a very significant role in many other important applications like designing of efficient burners, mixing in chemical reactors, thrust augmentation in aerospace applications, heat transfer etc. Most of the mixing and reaction takes place in the developing region of the jet of a combustor. Density differences between the jet and the ambient fluid play a major role in mixing. However, we will ignore these effects because of their complexity. Only the near field of an incompressible circular jet in coflowing streams is considered. The length of the potential core and the location of the virtual origin are integral parts of the near field-flow analysis and are important. That is why, we shall explore more about these two parameters, vis-a-vis their dependence on the nature of flow at the efflux section of the nozzle.

8.2 Potential Core

In the immediate downstream of the discharging nozzle, the shear layer spreads both inwards and outwards with reference to the axis of the jet. This is shown schematically in Fig 8.1. The edges spread almost linearly with the distance x. The distance from the nozzle at which the inner edge meets the jet axis is known as the length of the potential core and is denoted as x_0. The conical shaped region represents the potential core, wherein the velocity U_0 is uniform across the cross-section. In the potential core, the gradient of the mean velocity in the radial direction is negligible, and the level of turbulence is very small.

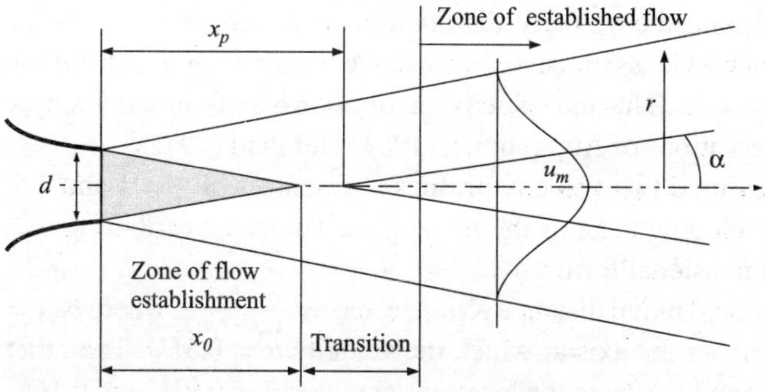

Figure 8.1 Definition sketch of the developing flow zone

It can be observed that for any location $x < x_0$, the velocity changes from a value of U_0 on the inner edge to 0, (or the ambient velocity U_a) at the outer edge. The lateral extent of the shear layer is small, therefore the velocity gradient in the shear layer is fairly large. For a coflowing jet, the rate of spread of the shear layer is affected by the velocity ratio $\beta \left(= \frac{U_a}{U_0}\right)$. Nonetheless, the features of the shear zone are identical to that of a jet in a stagnant ambient. The only difference is that for a coflowing jet, the velocity excess $(u - U_a)$ takes the role of the velocity u of a jet in a stagnant ambient. The potential core and the transition distance together constitute the 'developing zone'. The principle of self-similarity is applicable to the flow within the annular shear layer. The values of the radial distances r_1 and r_2 demarcate the extent of the shear layer at different x-locations. These can be found from experiments or through semi-empirical relationships.

It may be noted that the flow occurring at the end of the potential core is the initial condition for the analysis of the jet in the fully developed region. Therefore, a complete understanding of coflowing jets needs a proper analysis of the flow in the initial region of the jet. For a general coflowing round jet, the excess velocity $(u - U_a)$ within the shear layer is denoted by Δu. Similarly, the excess velocity of the core $(U_0 - U_a) = \Delta U_0$. The quantity Δu is zero at the outer edge $r = r_2$ and equals ΔU_0 at the inner edge $r = r_1$. The case in which $U_a = 0$ will be known as the simple annular shear layer.

8.2.1 Simple Annular Shear Layer

The simple annular shear layer was studied by many researchers, including Abramovich (1963) and Sami et al. (1967). Similarity analysis, as well as experimental observations, indicated that r_1 and r_2 vary linearly with the distance x. In addition, the velocity distribution in the shear layer can be represented by either a Gaussian curve, a Cosine expression or a distribution akin to the turbulent wake. That the velocity distribution is self-similar is amply clear from the observations of Abramovich (1963) and Pani (1972).

Abramovich measured two velocity profiles at distances of $\frac{x}{d} = 1$ and 2.5, well within the developing zone of the jet, and the velocity distribution was plotted in a non-dimensional form.

The non-dimensional radial distance was selected as $\frac{(r-b_{0.5})}{(r_{0.9}-r_{0.1})}$, where $b_{0.5} =$ the radial distance from the axis at which the velocity $u = 0.5U_0$. The other radial distances $r_{0.9}$ and $r_{0.1}$ were the locations at which $u = 0.9U_0$ and $0.1U_0$, respectively.

Half-velocity length scale

By following a more conventional approach, the half-width $b_{0.5}$ is selected as the length scale for the case of a coflowing jet. The observed distribution of the velocity could be plotted as

$$\frac{\Delta u}{\Delta U_0} = f\left(\frac{r - r_1}{b_{0.5}}\right) \tag{8.1}$$

Figure 8.2 shows the self-similarity exhibited by the measured velocity profiles for the simple annular shear layer, that is, $U_a = 0$. The observed data for non-zero values of U_a also resulted in similar plots (Pani, 1972).

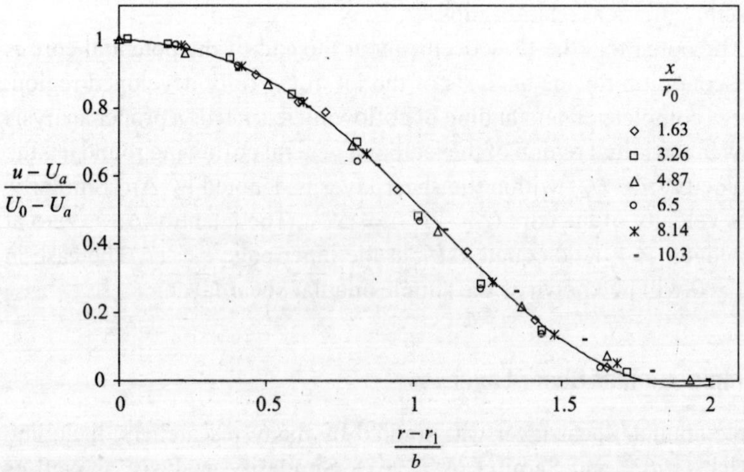

Figure 8.2 Self-similarity of the observed velocity: $U_a = 0$

To find how r_1 and r_2 (or $b_{0.5}$) vary with the distance x, we can write the integral momentum equation as

$$\frac{d}{dx}\int_0^{\infty} \rho r dr u \Delta u = 0 \tag{8.2}$$

which depicts that the excess momentum flux in the axial direction is preserved.

From the self-similarity property of the excess velocity distribution, we have

$$\frac{\Delta u}{\Delta U_0} = f\left(\frac{r - r_1}{b}\right) = f(\eta) \tag{8.3}$$

For the simple annular shear layer, the ambient velocity being zero, $\Delta u = u$ and $\Delta U_0 = U_0$. For such a case, the integral momentum equation can be rewritten as

$$\frac{d}{dx}\int_0^{r_1} \rho r dr U_0^2 + \frac{d}{dx}\int_{r_1}^{\infty} \rho r dr u^2 = 0 \tag{8.4}$$

First relationship between r_1 and r_0

Using the relationship $\frac{u}{U_0} = f(\eta)$ in Eq 8.4 and making use of the initial conditions $b = 0$ and $r_1 = r_0$ for $x = 0$, the following expression,

$$\left(\frac{r_1}{r_0}\right)^2 + \left(2 f_2 \frac{b}{r_0}\right)\frac{r_1}{r_0} + \left\{2 f_1 \left(\frac{b}{r_0}\right)^2 - 1\right\} = 0 \tag{8.5}$$

can be arrived at (Rajaratnam, 1976).

The solution of the above equation is

$$\frac{r_1}{r_0} = \sqrt{\left(\frac{b}{r_0}\right)^2 \{f_2^2 - 2 f_1\} + 1} - \left(\frac{b}{r_0} f_2\right) \tag{8.6}$$

where, $f_1 = \int_0^{\infty} \eta f^2 d\eta$ and $f_2 = \int_0^{\infty} f^2 d\eta$.

The observed distribution function $f(\eta)$ was used to evaluate the numerical values of f_1 and f_2 as 0.065 and 0.31, respectively. Using these values in Eq 8.6, one can write a simpler equation in the following form:

$$\frac{r_1}{r_0} = 1 - \left(\frac{b}{r_0}\right) f_2 \tag{8.7}$$

Second relationship between r_1 and r_0

The analysis based on self-similarity is not able to give the required second equation between r_1 and r_0. Therefore, setting up a second equation in r_1 and

r_0 (or b), a different approach was followed by integrating the equations of motion in the limits $r = 0$ to $\frac{(r_1+r_2)}{2}$ (Squire and Trouncer, 1944). This is elucidated in Section 8.3.

Alternatively, by assuming that the length scale varies linearly with the distance x, from the momentum integral equation the following was arrived at (Albertson et al., 1950):

$$\frac{r_1}{r_0} = 1 - \frac{x}{x_0} \tag{8.8}$$

Experimental data from different sources (Pani, 1972) showed this linear relationship holds good for all values of β. Figure 8.3 shows the data plotted from different sources.

Interestingly, the observed length of the potential core in a stagnant ambient is seen to be $4.9d$ and is significantly different from the commonly accepted value of $6.2d$. From Fig 8.3, the variation of r_1 can be expressed as

$$\frac{r_1}{r_0} = 0.95 - 0.097\frac{x}{r_0} \tag{8.9}$$

Again, based on the experimental results shown in Figs 8.4 and 8.5, the growth of the half velocity scale b and the spread of the outer edge of the

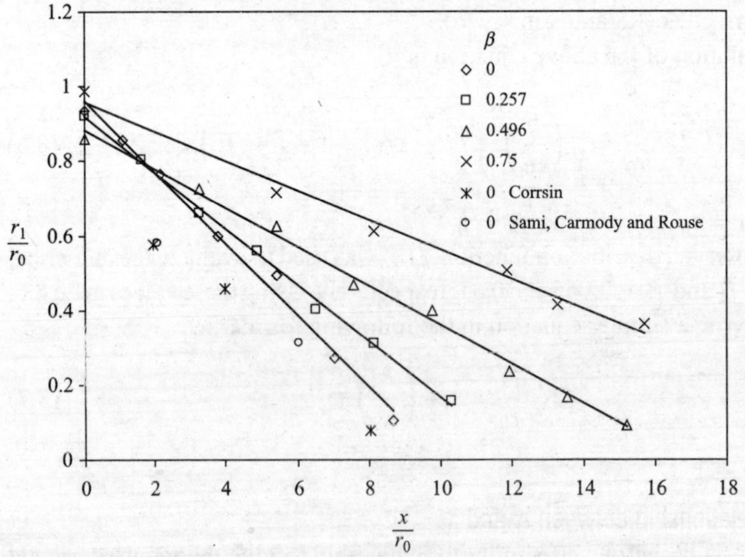

Figure 8.3 Observed variation of the inner edge of the shear layer

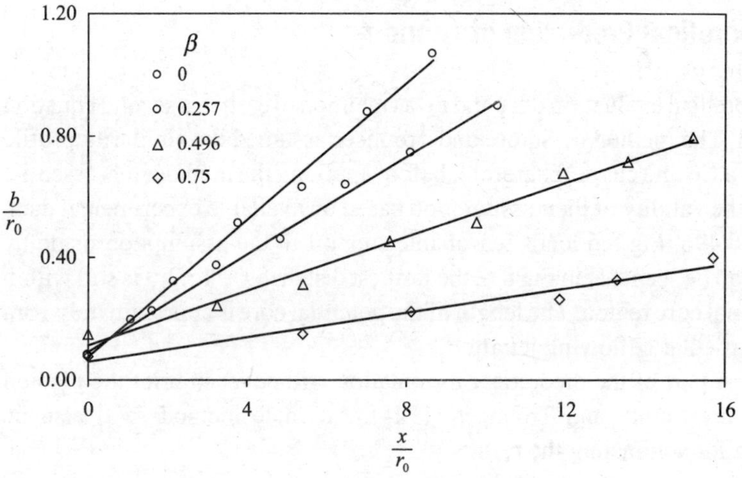

Figure 8.4 Growth of the length scale b with distance x

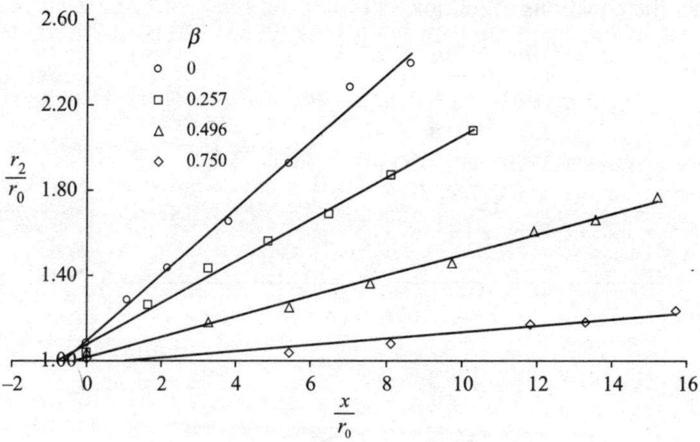

Figure 8.5 Spread of the outer edge r_2 with distance x

shear layer r_2 for the stagnant ambient case ($\beta = 0$) can be represented by the following linear expressions:

$$\frac{b}{r_0} = 0.10 + 0.111\frac{x}{r_0} \tag{8.10}$$

and

$$\frac{r_2}{r_0} = 1.07 + 0.158\frac{x}{r_0}. \tag{8.11}$$

8.3 Theoretical Prediction of r_1 and r_2

For a theoretical evaluation of r_1 and r_2, as mentioned earlier, a second equation is needed. The method of Squire and Trouncer assumed the similarity profile to follow a Cosine curve. Therefore, before applying the method, it is essential to check the validity of their assumption based on available experimental data. For $\beta = 0.496$, Fig 8.6 lends reasonable support to the assumption made by them. It can be seen that in Fig 8.6, the farthest distance $x = 15.2r_0$ is still within the potential core region. The length of the potential core is approximately $16r_0$ in this particular coflowing jet.

The first part of the theoretical formulation will cover in brief the method proposed by Squire and Trouncer (1944). Alternate methods will also be presented for contrasting the results.

8.3.1 Squire and Trouncer's Method

Let us consider the equations of motion

$$u\frac{\partial u}{\partial x} + v\frac{\partial u}{\partial r} = \frac{1}{\rho}\frac{1}{r}\frac{\partial (r\tau)}{\partial r} \tag{8.12}$$

and

$$\frac{\partial (ru)}{\partial x} + \frac{\partial (rv)}{\partial r} = 0 \tag{8.13}$$

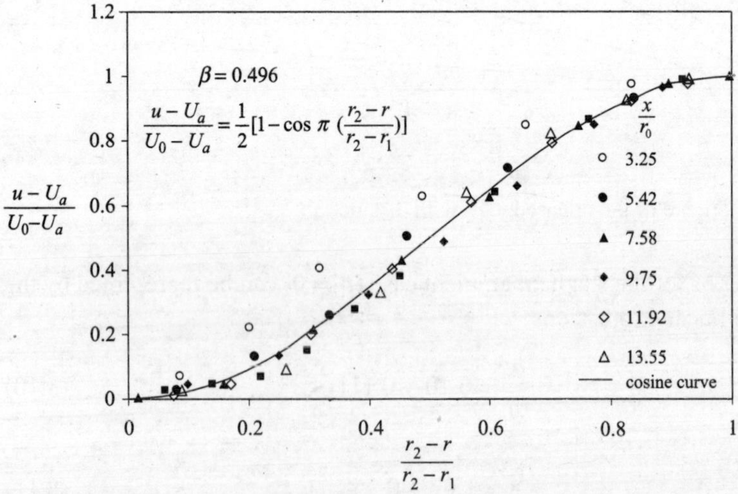

Figure 8.6 Cosine distribution of the excess velocity in the annular shear layer

Let us multiply the first equation by ρr and then integrate with respect to r. The resulting equation is

$$\frac{\partial}{\partial x}\int_0^r \rho u^2 r dr - \rho u \frac{\partial}{\partial x}\int_0^r u r dr = r\tau \tag{8.14}$$

The similarity profile was assumed to be a Cosine function

$$\frac{u - U_a}{U_0 - U_a} = \frac{\Delta u}{\Delta U_0} = \frac{1}{2}(1 - cos\pi\zeta) \tag{8.15}$$

where the non-dimensional distance $\zeta = \frac{(r_2 - r)}{(r_2 - r_1)}$.

Furthermore, the mixing length relationship of Prandtl was adopted in the form

$$\tau = \rho l^2 \left(\frac{\partial u}{\partial r}\right)^2 = \rho \{c(r_2 - r_1)\}^2 \left(\frac{\partial u}{\partial r}\right)^2 \tag{8.16}$$

A value of $c = 0.082$ was adopted for all values of the velocity ratio β, including the stagnant ambient case.

Equation 8.14 was evaluated for $r = r_2$ as well as for $r = \frac{(r_1 + r_2)}{2}$, which resulted in the following pair of equations,

$$a_{11} r_1^2 + 2a_{10} r_1 r_2 + a_{00} r_2^2 = r_0^2 \tag{8.17}$$

and

$$\frac{d}{dx}\{A_{11} r_1^2 + 2A_{10} r_1 r_2 + A_{00} r_2^2\} = B\{r_1 + r_2\} \tag{8.18}$$

where the coefficients $a_{11}, a_{10} \ldots A_{11}, A_{10}$ and B are functions of the velocity ratio β.

For example, $a_{11} = 2\left\{\frac{5-\beta}{16} - \frac{1}{\pi^2}\right\}$, $a_{10} = \frac{2}{\pi^2}$ etc.

Equations 8.17 and 8.18 were solved simultaneously, and the values of $\left(\frac{r_1}{r_0}\right)$ and $\left(\frac{r_2}{r_0}\right)$ were plotted as a function of the non-dimensional distance $\left(\frac{c^2 x}{r_0}\right)$ for different values of β, shown in Fig 8.7.

Simplified solutions

Pani (1972) did a similarity analysis and derived linear relationships between $r_1 - x$ and $r_2 - x$. That analysis simplified the procedure for solving the equations of Squire and Trouncer. A brief account of the method of evaluating the variables r_1 and r_2 is given below.

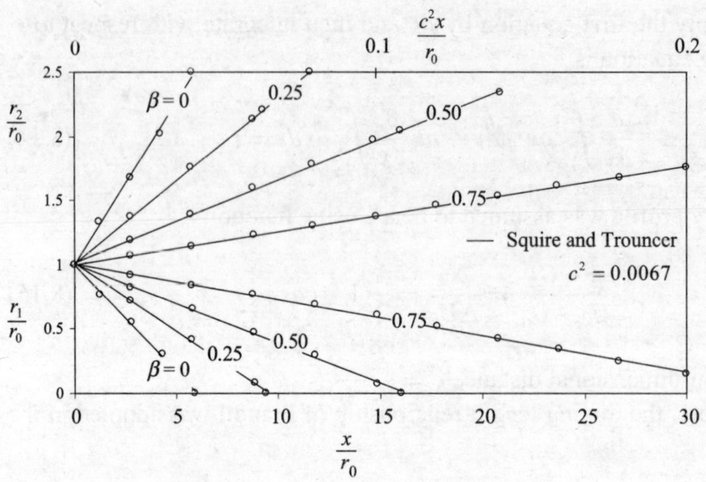

Figure 8.7 Theoretical values of r_1 and r_2 for different x

Let $R_1 = \frac{r_1}{r_0}$, $R_2 = \frac{r_2}{r_0}$ and $X = \frac{x}{r_0}$. In addition, let us write $R_1 = 1 - m_1 X$ and $R_2 = 1 + m_2 X$, where m_1 and m_2 are the unknown slopes which depend on β. Making use of the boundary conditions, $R_1 = 1$ and $R_2 = 1$ at $X = 0$, two algebraic equations were framed as an alternative to the original equations of Squire and Trouncer. The modified pair of equations is

$$a_{11} R_1^2 + 2a_{10} R_1 R_2 + a_{00} R_2^2 = 1 \tag{8.19}$$

and

$$A_{11} R_1^2 + 2A_{10} R_1 R_2 + A_{00} R_2^2 - R_1 \frac{BX}{2} - R_2 \frac{BX}{2} = BX + A_{11} + 2A_{10} + A_{00}. \tag{8.20}$$

The results of this procedure were found to be nearly identical to those of Squire and Trouncer. This is depicted in Fig 8.7 and gives credence to the proposed linear relationship between $r_1 - x$ and $r_2 - x$ for all the values of β.

8.3.2 Other Methods

Spreading hypothesis

Chu and Lee (1999) adopted the spreading hypothesis enumerated in Section 6.5.1 to establish a relationship for the potential core length in terms of the velocity ratio β. In the shear layer, the average velocity of the mixing layer was considered as $\frac{(U_0+U_a)}{2}$. Therefore, the average excess velocity $|\Delta U| = \frac{(U_0+U_a)}{2} - U_a = \frac{(U_0-U_a)}{2}$.

Using the principle of conservation of the excess momentum and the spreading hypothesis, the length of the potential core was established as

$$\frac{x_0}{d} = \frac{\sqrt{1+\beta}}{\beta_s(1-\beta)} \tag{8.21}$$

where β_s is the spread coefficient of a top-hat profile in a stagnant ambient. In a stagnant ambient, it has been observed that $\beta_s = 0.154$, therefore, the length of the potential core $x_0 = \frac{d}{\beta_s} = \frac{d}{0.154} = 6.5d$.

Point-source method

As an offshoot of Reichardt's hypothesis, the point-source technique can be adopted to predict the maximum velocity on the axis of the circular jet in the developing region of flow. The length of the potential core can be defined as that distance downstream of the nozzzle at which the maximum excess momentum $u_m \Delta u_m$ is $0.99 U_0 \Delta U_0$. In the annular shear layer surrounding the potential core, the excess momentum flux satisfies similarity. Since the ratio of the maximum excess velocity u_m and the ambient velocity U_a is constant for $x \leq x_0$, the spread coefficient c in the relationship $b = cx$ remains the same at every section of the shear layer.

Based on Reichardt's hypothesis, the excess momentum flux on the axis of an elemental area dA was found in Chapter 6 as

$$\frac{u \Delta u}{U_0 \Delta U_0} = \frac{dA}{2\pi b^2} e^{-\left(\frac{\xi^2}{2}\right)} \tag{8.22}$$

The governing equation of motion being linear in $u\Delta u$, the value of the excess momentum flux on the axis of the jet can be obtained by the method of superposition. Therefore, on the axis of the circular jet, the excess momentum flux can be shown to be

$$\frac{u_m \Delta u_m}{U_0 \Delta U_0} = 1 - \exp\left\{-\frac{1}{2}\left(\frac{d}{2cx}\right)^2\right\} \tag{8.23}$$

Interestingly, the above relationship holds good for the developing as well as the developed region of the flow. Therefore, it is not necessary to artificially divide the flow into various zones. The transition between different zones takes place in a seamless fashion.

Imposing the forementioned 99 per cent criterion from Eq 8.23 we can arrive at the expression

$$\frac{x_0}{d} = \frac{0.165}{c} \tag{8.24}$$

In a stagnant ambient, the value of the spread coefficient c, based on Reichardt's hypothesis, is approximately equal to 0.056. Hence, the length of the potential core in a stagnant ambient $x_0 = 2.94d$.

Empirical expression

Forstall and Shapiro (1950) proposed an empirical expression for the potential core length as a function of the velocity ratio as

$$\frac{x_0}{d} = 4 + 12\beta \tag{8.25}$$

Therefore, for a jet in a stagnant ambient, the length of the potential core $x_0 = 4d$.

8.4 Potential Core Observed in Experiments

The velocity decay relationship is often written as $\frac{u_m}{U_0} = \frac{C_u}{\frac{(x-x_p)}{d}}$, where x_p is the distance of the virtual origin (to be discussed later in this chapter) from the face of the nozzle, and C_u is a constant. In the far field region, the distance x is much larger than x_p. Therefore, the role of the virtual origin is quite insignificant, and to find the value C_u from experimental data, $\frac{U_0}{u_m}$ is plotted against $\frac{x}{d}$. The slope of the best-fit linear relationship gives the value of C_u. Alternatively, the ratio $\frac{u_m}{U_0}$ is plotted against $\frac{x}{d}$, and the curve is extrapolated backward. The largest value of $\frac{x}{d}$ at which $\frac{u_m}{U_0} = 1$ marked by an arrow in Fig 8.8 also gives the value of C_u. The value is considered as the representative (hypothetical) length of the potential core, $\frac{x_0}{d}$.

It is expected that this artificially generated length will not match with the actual distance at which $u_m \sim U_0$ is satisfied. This is clear from the plot shown in Fig 8.8. At locations very close to the nozzle, $4 \leq \frac{x}{d} \leq 12$, the observed values of the maximum velocity follow a different trend. Experimental data from independent sources, shown in the figure, indicate that the real length of the potential core is close to $4d$ and agrees with the empirical value suggested in Eq 8.25. The length of the potential core was also measured directly for different β and for $\beta = 0$, x_0 was equal to $4.9d$ (Pani, 1972).

The length of the potential core predicted by extrapolating the velocity decay curve backward does not exist physically. For a stagnant ambient, this length lies in the range of $5.8d$ to $6.2d$ based on the data of different experiments. On an average it can be assumed that $x_0 = 6d$. If the virtual origin correction is taken into account, the value of C_u, which is also the value of $\frac{x_0}{d}$, lies between 5.7 to 6.06 (Malmström et al., 1997).

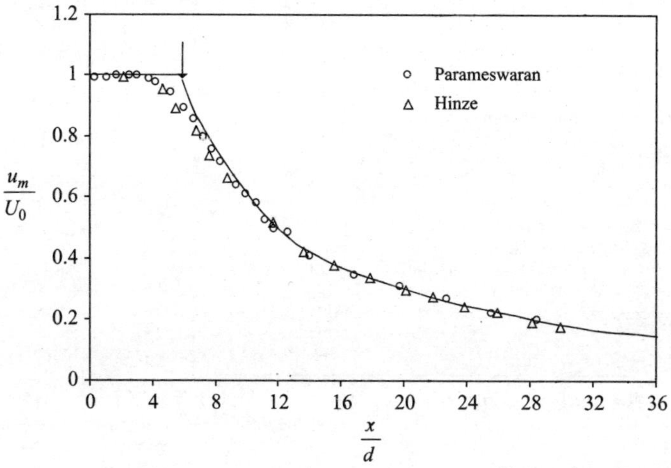

Figure 8.8 Velocity decay in the proximity of the circular nozzle

Table 8.1 Length of potential core in a coflowing stream

$\beta\left(=\frac{U_a}{U_0}\right)$	$\frac{x_0}{d}$ (observed)	Squire and Trouncer	Forstall and Shapiro	Chu and Lee	Point source
0	4.9	3.0	4.0	6.5	3.05
0.257	6.1	4.8	7.08	9.8	5.17
0.496	8.15	8.1	9.95	15.7	9.07
0.750	12.5	19.3	13.0	34.3	21.38

When $\beta \neq 0$, different methods to determine the length of the potential core in a coflowing stream, yield different answers for x_0. By assuming $c = 0.0818$ in the mixing length hypothesis, the length of the potential core based on Squire and Trouncer's approach is evaluated for four different values of β and the results are shown in Table 8.1.

The same four velocity ratios were used in the experiments for finding the actual distance at which the inner edge of the shear layer intersected the axis of the jet. The observed data are also shown in the above Table. For a meaningful comparison, the value of c in Squire and Trouncer's solutions for different β was evaluated by matching their theoretical results with observations (Pani, 1972). The coefficient c was found to vary with β and is not a single-valued entity. The coefficient takes the values of 0.064, 0.072, 0.081 and 0.098 for $\beta = 0$, 0.257, 0.496 and 0.75, respectively.

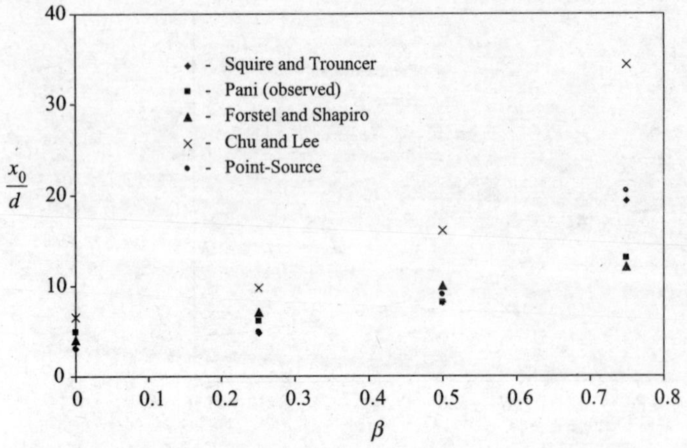

Figure 8.9 Effect of ambient velocity on the length of the potential core (Source: Pani, 2008)

Table 8.1 compares the length of the potential core obtained from different methods. It is evident that Chu and Lee's method yields much larger values for $\frac{x_0}{d}$ compared to those of others. In applying the point-source technique, the spread coefficient in a stagnant ambient is considered $c_{\text{stag}} = 0.054$. For non-zero values of β, the general spreading hypothesis can be adopted for the coefficient c as

$$c = (c_{\text{stag}})\frac{1-\beta}{1+\beta} \tag{8.26}$$

The length of the potential core corresponding to the aforesaid β values was evaluated from Eq 8.24. Figure 8.9 shows the variation of $\frac{x_0}{d}$ with the velocity ratio β, based on values obtained from different methods.

No single method is capable of predicting the length of the potential core satisfactorily for all the values of β considered here. *The point source method is able to show the gradual variation in the value of u_m in the proximity of the nozzle and looks more realistic. It obliterates the artificial imposition of a developing and a developed zone downstream of the nozzle.* Forstall and Shapiro's empirical equation seems to be in fair agreement with the experimental results of Pani (1972). For the highest value of $\beta = 0.75$, the predictions are not as good. Nonetheless, it is important to interpret the oft quoted $x_0 = 6.2d$ as a hypothetical length of the potential core in a stagnant ambient. The real length of the core happens to be much smaller.

Next, let us examine another important feature of the flow, the 'virtual origin'.

8.5 Virtual Origin

For a circular jet issuing from a nozzle of diameter d, the momentum flux $M_0 = \rho \frac{\pi}{4} d^2 U_0^2$. It has been shown that for a zero-pressure gradient flow, the momentum flux is conserved (section 3.4) and at any location x, the momentum flux $M = M_0$. The momentum flux M_0 solely governs the behaviour of the jet if the distance x is measured from a fictitious point called the virtual origin. The individual parameters like the diameter and the efflux velocity are of no consequence.

As the relationship between the length scale b and the distance x is linear, the similarity distribution in the fully developed region of the jet can be expressed as $\frac{u}{u_m} = f\left(\frac{r}{x}\right)$. The equal velocity lines $\frac{u}{u_m} = 0.9, 0.7, 0.5$ etc., will appear as rays and will intersect at a point on the x-axis, the pole of the jet. Experimental data from different sources do support this inference. Along the ray $\frac{u}{u_m} = 0.5$, the relationship $r = b_{0.5}$ is satisfied, and the observed growth rate $\frac{b_{0.5}}{x} = 0.097$ (Abramovich, 1963).

Almost all the maximum velocity decay relationships that have been presented in this book reckon the distance x from the virtual origin (or the pole) of the jet. The location of the pole is not unique and the parameters that affect it will be presented herein. To avoid uncertainties, normally the distances are measured from the face of the nozzle and minor modifications are incorporated either in the decay constant or the exponents of the distance x. These are followed, especially, for three-dimensional free jets.

The virtual origin based on the length scale b seldom coincides with the location found from velocity decay relationship. It has been established experimentally that the location of the virtual origin is quite sensitive to the turbulence level in the nozzle and not so sensitive to the nozzle design or details (Flora Jr. and Goldschmidt, 1969). It was found that the virtual origin from velocity considerations was generally located behind the face of the nozzle.

Measurements of Wygnanski and Fiedler (1969) show that $x_p \sim 3d$ if the measured points lie within $x \simeq 50d$. Hussein et al., (1994) measured velocities in a high-Reynolds-number circular jet, and based on the data points that are within $x \simeq 100d$, the virtual origin was located at a distance of $4d$, downstream of the outlet. The values of the decay constant, C_u and the position of the virtual origin from different experimental investigations are summarised in the following Table 8.2.

The existence of a negative virtual origin, observed in some experiments, has been attributed to the occurrence of backflow at the origin when $U_a = 0$.

Table 8.2 Decay constant and position of virtual origin

Authors	C_u	$\frac{x_p}{d}$
Wygnanski & Fiedler (1969)	5.7	3.0
Panchapakesan & Lumley (1993)	6.06	−2.5
Hussein et al. (1994) LDA	5.8	4.0

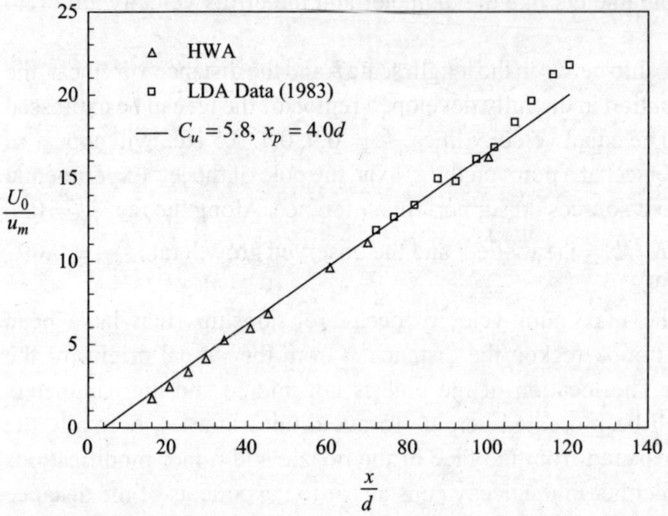

Figure 8.10 Location of the virtual origin for a circular jet (Source: Hussein et al., 1994)

Figure 8.10 shows the centreline velocity variation from different experiments resulting in $x_p \simeq 4d$ for a stagnant ambient.

In contrast to the observations of Flora and Goldschmidt, Mesbah Uddin and Pollard (2007) have concluded that the turbulence intensity at the orifice plays no significant role in the determination of the virtual origin location. According to them, the velocity distribution at the jet inlet, characterized by the local value of the variance in the boundary layer thickness, affects the location of the virtual origin. For locating the virtual origin, they provide a theoretical expression in terms of β and for $\beta \to 0$, $\frac{x_p}{d} \simeq 4$.

Malmströrm et al. (1997) conducted experiments on low-velocity axisymmetric air jets. It was observed that $\frac{x_p}{d}$ was between 3 to 4 for higher velocities, decreases to 0 or −1 for velocity of about 6 m/s and then increases when the outlet velocity is further decreased. This behaviour of the jet is shown

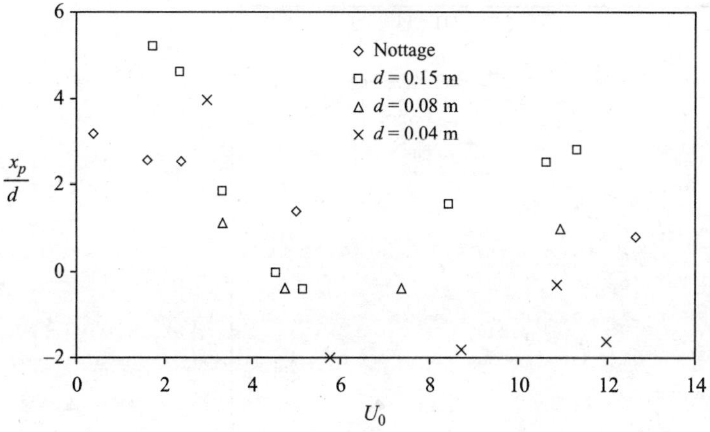

Figure 8.11 Variation in the position of the virtual origin for low-velocity jets (Source: Malmströrm et al., 1997)

in Fig 8.11. The new decay constant C_u for different sizes of the nozzle was best correlated by the nozzle outlet velocity. A co-variation between C_u and x_p values seems to exist. It is worth reiterating the fact that the jet exit conditions not only influence the velocity decay relationship but also the temperature decay. In the recent investigations by Mi et al. (2001), two types of outlets such as a smooth contraction and a long straight pipe were used to measure the temperature field for a Reynolds number of 16,000. Identical test facilities were adopted and measurements extended from near the outlets to a distance of 70d downstream. Some important conclusions arising out of the study are as follows:

(a) the centreline decay rate of the mean temperature is larger for the jet issuing from a smooth contraction than for a long pipe. This is true for the near as well as the far field.
(b) The presence of a single virtual origin could not be established.
(c) Most importantly, differences in the temperature field were observed even in the self-similar region. Thus, the entire flow is influenced by the initial condition. Hence, it is unlikely that an universal asymptotic state of turbulence can exist for different classes of turbulent flows. The earlier inference that the jet in the fully developed state loses memory may need a relook.

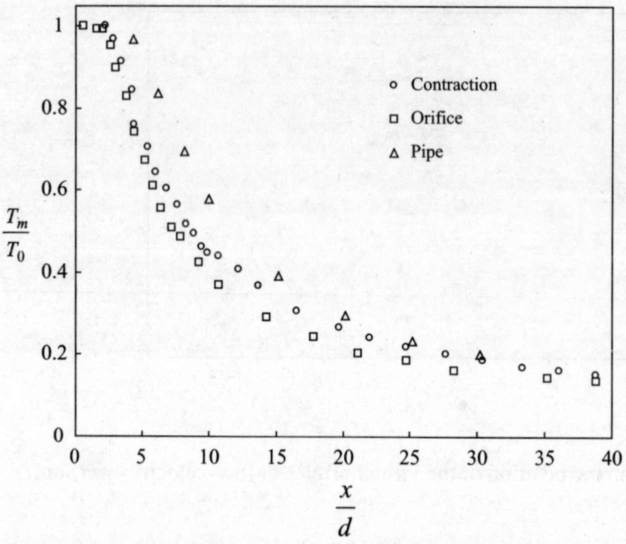

Figure 8.12 Normalised centreline temperature variation for different outlet conditions (Source: Mi et al., 2001)

Figure 8.12 shows the normalised mean centreline temperature variation for different types of outlets. Our earlier observation on the role of the conditions at the jet inlet can be clearly seen. The length of the potential core is seen to be the largest for the pipe-flow case, while there is a forward shift in the location of the virtual origin. The way the temperature half-widths vary with the distance x for different jet inlet conditions is shown in Fig 8.13 based on the measurements of Mi et al. (2001). It may be noted that the locations of the virtual origin are different from the ones based on the velocity decay relationship.

It has been observed that the width of the circular jet varies linearly with the distance x, whereas the maximum velocity u_m decreases linearly with the distance x. This is true when the distances are measured from the virtual origin located at x_p. Therefore, a useful relationship for predicting the dilution achieved at various downstream distances can be written as,

$$\frac{Q}{Q_0} = K_v \frac{(x - x_p)}{d} \tag{8.27}$$

where K_v value lies between 0.33 and 0.36. Equation 8.27 can be recast to express $\frac{Q}{dv}$ as a linear function of the Reynolds number R_e. By accounting for the location of the virtual origin, the results from different sources can be made to fall on a single line, as depicted in Fig 8.14.

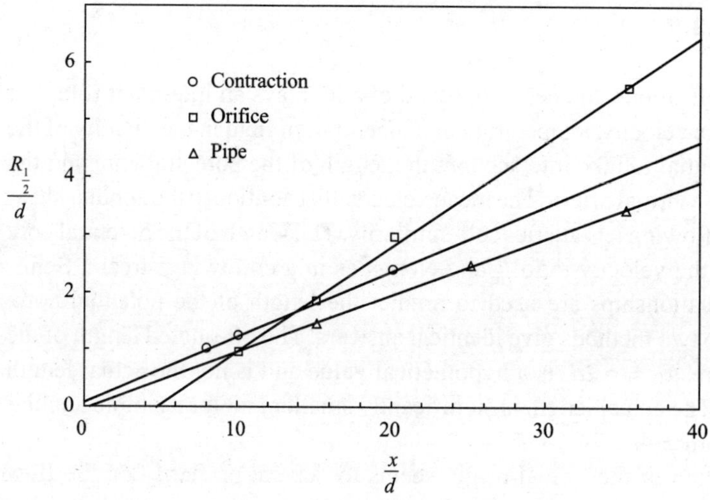

Figure 8.13 Temperature half-width for different jet inlet conditions

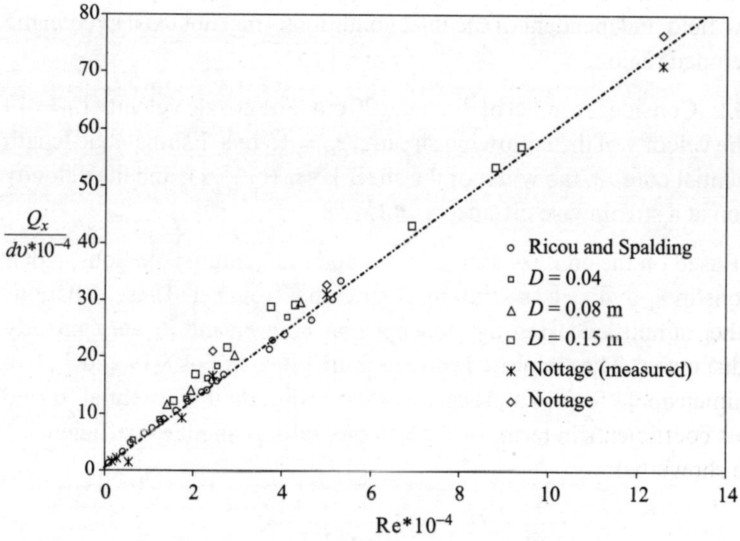

Figure 8.14 Q at $\frac{x}{d} = 25.6$ versus Reynolds number (source: Malmström et al., 1997)

8.6 Summary

In many applications, the near flow field of a jet plays an important role. For predicting the velocity, temperature or concentration field in the vicinity of the jet, it is essential to take into account the length of the potential core and the location of the virtual origin. The mean velocity distribution in the annular shear layer of a coflowing jet satisfies self-similarity. The length of the potential core depends on the velocity ratio β and elongates in a coflowing stream. Semi-empirical relationships are used to predict the length of the potential core. However, no two methods give identical answers. The oft-quoted length of the potential core, $x_0 = 6.2d$, is a hypothetical value and is not the actual length of the core. The actual length is significantly smaller and for a simple annular shear layer, $x_0 \simeq 4d$.

The location of the virtual origin seems to depend primarily on the flow conditions that prevail at the jet inlet. Generally, the virtual origin is at a downstream distance of 2 to 4 times the diameter of the nozzle. There are instances when it has been located behind the nozzle. A reliable estimate of the location of the virtual origin, based on theory, is difficult to find. Low-velocity jets have a larger variation in the location of the virtual origin. Self-similarity of the flow field, independent of the inlet conditions, may not exist even in the fully developed region.

Example 8.1: Consider an air jet of diameter 20 cm. The nozzle velocity $U_0 = 40$ m/s and the velocity of the coflowing stream, $U_a = 15$ m/s. Estimate the length of the potential core x_0, the width of the shear layer $(r_2 - r_1)$, and the velocity distribution at a streamwise distance $x = 1.0$ m.

Solution: Based on the conservation of mass and momentum equations, a pair of equations in r_1 and r_2 was setup by Squire and Trouncer. These equations were further simplified using the concept that both r_1 and r_2 vary linearly with the distance x. The simplified equations are given in Eqs 8.19 and 8.20 in the non-dimensional form. To proceed with the solution it is essential to find the various coefficients in terms of the velocity ratio β and the coefficient c^2. These are shown below.

$$\beta = \frac{U_a}{U_0} = \frac{15}{40} = 0.375$$

For a velocity ratio equal to 0.375, the corresponding value of $c^2 = 0.0058$. Hence,

$$a_{00} = 2\left\{\frac{3+\beta}{16} - \frac{1}{\pi^2}\right\} = 0.2192$$

$$a_{10} = \frac{2}{\pi^2} = 0.2026$$

$$a_{11} = 2\left\{\frac{5-\beta}{16} - \frac{1}{\pi^2}\right\} = 0.3755$$

$$A_{00} = \left\{\frac{1-\beta}{16} + \frac{1+\beta}{2\pi} - \frac{5+3\beta}{4\beta^2}\right\} = 0.10275$$

$$A_{10} = \left\{\frac{1-\beta}{16} + \frac{5+3\beta}{4\pi^2}\right\} = 0.1942$$

$$A_{11} = \left\{\frac{13+3\beta}{16} - \frac{1+\beta}{2\pi} - \frac{5+3\beta}{4\pi^2}\right\} = 0.5088$$

$$B = -\frac{\pi^2}{2}\{1-\beta\}c^2 = -0.01789$$

With the known values of the coefficients, Eqs 8.19 and 8.20 can be solved simultaneously using an iterative technique. This was followed in the past. However, there is an alternate simple procedure which can yield the necessary answers, that is, R_1 and R_2. This is illustrated below.

At the end of the potential core, $R_1 = 0$.

Hence, from Eq 8.19, $R_2 = \sqrt{\frac{1}{a_{00}}} = \frac{1}{\sqrt{0.2192}} = 2.136$

Substituting $R_1 = 0$ in Eq 8.20, we have

$$A_{00}R_2^2 - R_2\frac{BX_0}{2} = BX_0 + A_{11} + 2A_{10} + A_{00}$$

where X_0 is the length of the potential core.

Thus, the length of the potential core, $X_0 = \frac{\left\{\frac{A_{00}}{a_{00}} - (A_{11}+2A_{10}+A_{00})\right\}}{B\left\{1+\frac{1}{2\sqrt{a_{00}}}\right\}}$

It is pertinent to notice that from the definitions of the coefficients, the sum $(A_{11} + 2A_{10} + A_{00}) = 1$.

$$\text{Hence, } X_0 = \frac{\left\{\frac{A00}{a_{00}} - 1\right\}}{B\left\{1+\frac{1}{2\sqrt{a_{00}}}\right\}}$$

An interesting feature of this expression is that to find the length of the potential core, one does not need all the coefficients that appear in the pair of simultaneous equations 8.19 and 8.20.

On substitution, we can get $X_0 = 14.36$.

Let us denote the R_2 value corresponding to $X = X_0$ as R_{20}. Since both R_1 and R_2 vary linearly with the distance X, we can write for $0 \leq X \leq X_0$,

$$R_2 = 1 + (R_{20} - 1)\left(\frac{X}{X_0}\right)$$

and

$$R_1 = 1 - \left(\frac{X}{X_0}\right)$$

Thus, at $X = \frac{100}{10} = 10$, $R_2 = 1.791$ and $R_1 = 0.303$. Hence, the outer and the inner edges of the shear layer are at distances of $r_2 = 1.791 \times r_0 = 0.1791$ m and $r_1 = 0.303 \times r_0 = 0.1488$ m, respectively.

The thickness of the shear layer, $(r_2 - r_1) = 0.1488$ m.

At $X = 10$, the excess velocity distribution is as follows:

$$\frac{\Delta u}{\Delta U_0} = \frac{1}{2}\left\{1 - cos\pi \frac{(r_2 - r)}{(r_2 - r_1)}\right\}$$

$$\text{or} \quad \Delta u = \frac{\Delta U_0}{2}\left\{1 - cos\pi\left(\frac{0.1791 - r}{0.1488}\right)\right\};\ 0.0303 \leq r \leq 0.1791 \text{ m}$$

where $\Delta U_0 = (U_0 - U_a) = 25.0$ m/s.

9 Energy Loss at Sewer Junction Boxes

9.1 Introduction

Access holes (manholes) are essential in a sewer system for the purpose of inspection, maintenance or repairing a segment of the drainage. These openings are usually provided at intervals of approximately 90–180 m along a given sewer and are circular or square in plan. In addition, chambers are also provided at every junction between multiple pipes. The junctions allow the pipes to change one or more of the variables like the elevation of the invert, diameter, orientation and the slope.

It has been noticed that the losses associated with the junction boxes play a critical role in designing a drainage network. Often, based on observations from physical models, formulae have been evolved to estimate the losses. Theoretical determination of the losses is difficult because of the very complex flow prevailing in the chamber, even in the case of the simplest layout, that is, the straight-flow through junctions. The conditions become even more intractable if the junction box has to cater to plunging flows, supercritical flows, benching and arbitrary inflow angles. Methods to take care of these complexities have been proposed recently in FHWA Urban Drainage Design Manual (2009).

As a matter of convention, various types of losses taking place in the junction box are expressed as an empirical coefficient multiplied by the velocity head in the outflow pipe. It is also common to assume that the pipes are flowing full, but not under pressure. In case of pipes, flowing full, under pressure, there will be a more stable flow condition, higher discharge, higher loss of energy and a fluctuating free surface in the box. When the depth of flow in the junction box is small, swirling and oscillating flows normally occur, causing larger loss of energy.

The inlet pipe suffers losses at its exit and the loss coefficient $K_i \simeq 0.4$. Similarly, at the outlet pipe (entrance), the loss coefficient K_o is approximately equal to 0.16. Experiments on a level inflow pipe and a outflow pipe, with the

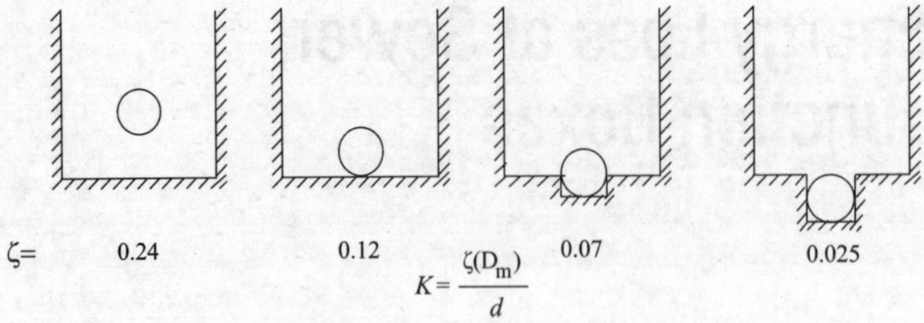

Figure 9.1 Head loss coefficient for different moulds

flow leaving the box without change in its orientation, have yielded a total-loss coefficient $K \sim 0.60$. Benching seems to have a very significant effect and leads to a reduction in the losses. Based on his own and others experiments, Marsalek (1984) expressed the head-loss coefficient K in terms of a shape factor ζ, and the relative width of the access hole chamber. The relationship is, $K = \zeta\left(\frac{D_m}{d}\right)$, for $\frac{D_m}{d} \leq 4$, where D_m is the width of the chamber.

In Fig 9.1, the proposed shape factor values for different moulds are given. It can be observed that the total-loss coefficient K decreases with a decrease in the size of the access hole chamber. The losses are considerably less compared to the conventional, combined expansion–contraction loss one would be tempted to use. This happens because the main body of the stream crossing the junction box remains more or less intact. Only the outlying parts undergo shearing and change in their trajectory. Benching produces the lowest loss and the reason for it is similar to the aforementioned one.

Sometimes, the inflow and outflow pipes are oriented at 90° and the head-loss in the junction box is significantly larger. Yoon et al. (2010) have summarised the results of many earlier investigators. For a surcharged flow, the shape of the junction box had very little effect on the head-loss coefficient. The loss coefficient was $1.6 \sim 1.7$ for both circular and square shapes of the chamber. Even the location of the pipes had a minor effect on the loss coefficient. For the case in which the two pipes were located at the corners of the square box, side benching helped to reduce the head loss coefficient by almost 30 per cent.

It will be demonstrated that the losses in the junction box, pertaining to the circulatory flow pattern within the chamber, can be predicted theoretically. The layout is the simple 180° outlet pipe orientation and the box is square in plan. The application of the free-jet theory, including the point-source technique of superposition, has been adopted in the past (Pederson and Mark, 1990, and Mudgal and Pani, 1993) with a fair amount of success. The following section gives a brief account of the methodology for predicting the head-loss coefficient of a junction box based on free-jet theory.

9.2 Free Jet in a Junction Box

In developing a theoretical expression for evaluation of the energy loss taking place in the junction box, the incoming liquid is assumed to behave like a classical, submerged free jet. The ambient water gets entrained by the jet and the forward flow increases in the streamwise direction. For a steady state condition, the inflow Q_1 must equal the outflow Q_2. So, the surplus water in the forward moving jet gets ejected from the main flow before leaving the chamber. A part of the energised ambient fluid, entrained in the beginning, is left behind in the manhole chamber. The energy loss taking place in the junction box can be attributed to this persistent pumping mechanism, that is, removal of some energised fluid from the forward moving jet.

The flow with depleted energy enters the outlet pipe and an additional loss of energy takes place downstream of the vena-contracta as given below:

$$\Delta h_v = K_v \left(\frac{U_2^2}{2g} \right) \tag{9.1}$$

The loss coefficient K_v is associated with the flow expansion that takes place after the vena-contracta in the outflow pipe. It is normally expressed as

$$K_v = 0.5 \left(1 - \frac{A_2}{A_{\mathrm{mj}}} \right) \tag{9.2}$$

In the above expression, A_2 is the cross-sectional area of the outflow pipe, and A_{mj} is the characteristic cross-sectional area of the manhole jet containing the initial discharge Q_1. For practical purposes, it is desirable to combine the loss taking place in the chamber, Δh_m, and the loss due to the formation of the vena-contracta in the outflow pipe, Δh_v. As the losses in the converging flow near the mouth of the outflow pipe is negligible, $K_v \simeq K_o$.

The ratio of the longitudinal length of the junction box, D_m, to the inlet pipe diameter d_1 is much less than 6. Therefore, any cross section of the jet located at a distance of x from the efflux section comprises of two zones, the potential core and the annular shear layer as shown in Fig 9.2. The total flux of volume and energy will be the sum of the discharges and energy contributed by the two zones.

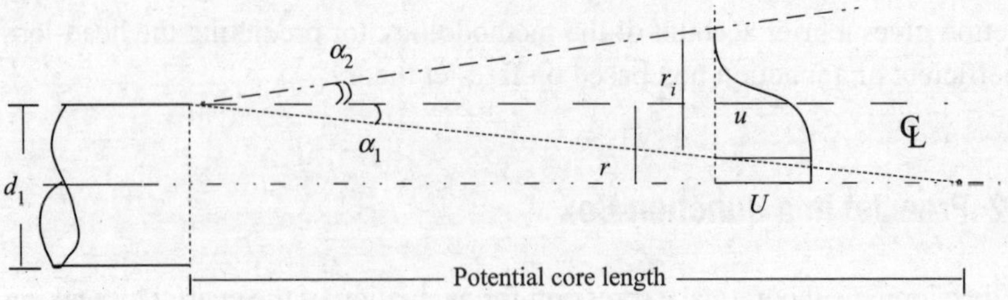

Figure 9.2 Definition sketch of potential core region

9.2.1 Contribution of Potential Core

The diameter of the potential core decreases linearly with the distance from the efflux section of the inflow pipe, Fig 9.2. Based on the observation that the velocity in the core is equal to U_1, the following relationships can be written,

$$\frac{Q_c}{Q_1} = \left\{1 - 2\left(\frac{x}{d_1}\right)\tan(\alpha_1)\right\}^2 \tag{9.3}$$

and

$$\frac{E_c}{E_1} = \left\{1 - 2\left(\frac{x}{d_1}\right)\tan(\alpha_1)\right\}^2 \tag{9.4}$$

In Eqs 9.3 and 9.4, $\alpha_1 (\simeq 4.6°)$ is the angle the inner edge of the shear layer makes with the axis of the jet, Q_c is the flow in the potential core, and E_c is the energy content of the flow in the potential core.

9.2.2 Shear Layer Contribution

As the outer edge of the shear layer expands in a linear fashion at an angle α_2 ($\simeq 9^0$), shown in Fig 9.2, it is obvious that the total thickness of the shear layer grows in a linear fashion with the distance x. Further, the velocity profiles in the shear layer are approximated by the Gaussian distribution. With the use of the assumed velocity distribution, the discharge and the energy flux in the shear layer can be obtained by integration. The limits of integration for the radial distance are the inner and the outer boundaries of the shear layer. Pederson and Mark (1990) have derived the required expressions to find $\frac{Q_s}{Q_1}$ and the kinetic energy flux $\frac{E_s}{F_1}$, where Q_s and E_s are the flow and energy flux in the shear layer, respectively. *We shall designate such an approach as the potential-core method to distinguish it from the point-source technique.* In the potential-core method, at any distance x, the discharge Q_{mj} and the kinetic energy flux E_{mj}

of the manhole jet can be evaluated from the following:

$$\frac{Q_{\mathrm{mj}}}{Q_1} = \frac{Q_c}{Q_1} + \frac{Q_s}{Q_1} \tag{9.5}$$

$$\frac{E_{\mathrm{mj}}}{E_1} = \frac{E_c}{E_1} + \frac{E_s}{E_1} \tag{9.6}$$

However, at $x = D_m$ the total forward flow entering the outflow pipe must equal the inflow discharge Q_1. Thus,

$$\frac{Q_s}{Q_1} + \frac{Q_c}{Q_1} = 1. \tag{9.7}$$

The radial distance of the jet boundary which satisfies Eq 9.7 will be known as the radius of the manhole jet, r_{mj}. The energy loss in the manhole is therefore

$$\Delta E_m = E_1 - E_2 = E_1 - \{E_{\mathrm{mj}}\}_{r=r_{\mathrm{mj}}} \tag{9.8}$$

where

$$E_1 = \gamma Q_1 \left(\frac{U_1^2}{2g}\right) \tag{9.9}$$

The corresponding head loss in the manhole is

$$\Delta h_m = \frac{\Delta E_m}{\gamma Q_1} \tag{9.10}$$

From Eqs 9.9 and 9.10, one can write the following relationship:

$$\Delta h_m = \left(\frac{\Delta E_m}{E_1}\right)\frac{U_1^2}{2g} = \left(1 - \frac{E_2}{E_1}\right)\frac{U_1^2}{2g} \tag{9.11}$$

Thus, from Eq 9.11, one can see that the coefficient for the head loss in the manhole chamber is $K_m = \left(1 - \frac{E_2}{E_1}\right)$. It may be noted that the the exit loss of the inlet pipe K_e is equal to the loss coefficient K_m. Consequently, the total head loss coefficient of the junction box

$$K = K_i + K_o = K_m + K_v \tag{9.12}$$

To find the total head loss coefficient theoretically, the loss in the junction box can be evaluated by either of the two approaches, namely the potential-core method or the point-source technique. The second approach will be presented here in some detail, and the ensuing results will be contrasted with those from the potential-core method (Pederson and Mark, 1990).

9.2.3 Head Loss Using Point-Source Method

The justification for superposing the square of the velocities (i.e., specific momentum flux) is given in Chapter 4. The cross-sectional area of the pipe A_1 is divided into a number of elemental areas, designated as δA_1. Point-sources having a momentum flux of $\rho U_1^2 \delta A_1$ are imagined to be located at the centroid of each of the elemental areas. Subsequent steps to be followed in the evaluation of the energy loss ΔE_m are as follows:

(i) A trial value for the target jet area A_j is assumed and is discretised into a number of equal-sized elemental areas, δA_j.
(ii) The co-ordinates of each elemental area, (x, r, θ), are determined.
(iii) The momentum flux across any elemental area δA_j is the sum total of the contributions from all the elemental source jets, δA_1's located at $x = 0$. Hence, at any selected point over the cross section of the jet, add

$$\frac{u^2}{U_1^2} = \frac{\delta A_1}{2\pi C^2 x^2} \exp\left\{-\left(\frac{r}{\sqrt{2}Cx}\right)^2\right\} \tag{9.13}$$

to account for all the δA_1s.
(iv) The total momentum flux across A_j is the sum of the momentum flux of the elemental areas δA_j, found in step (iii).
(v) On similar lines, the volume flux across each δA_j is added to find the discharge Q_j.
(vi) If the discharge computed in step (v) is equal to ±1 per cent of Q_1, then the trial value of $A_j = A_{mj}$, else steps (i) to (vi) are repeated with a new trial value for A_j.

For the final value of A_j, which represents A_{mj}, the momentum flux is equal to the value found in step (iv). The energy flux E_{mj} is calculated for the flow across A_{mj} and the loss of energy in the junction box is found from Eq 9.8. It is important to note that the above procedure does not distinguish between the zone of flow establishment and the developed flow region. Further, it only needs a single empirical coefficient C to make it operational.

9.3 Head Loss Coefficient Results

Following the potential-core approach, analytical expressions were given by Pederson and Mark (1990) and Abramovich (1963) for finding the momentum and the energy flux contained in the core of the initial flow of an axially symmetric jet. These relationships allow to find the variation of the momentum

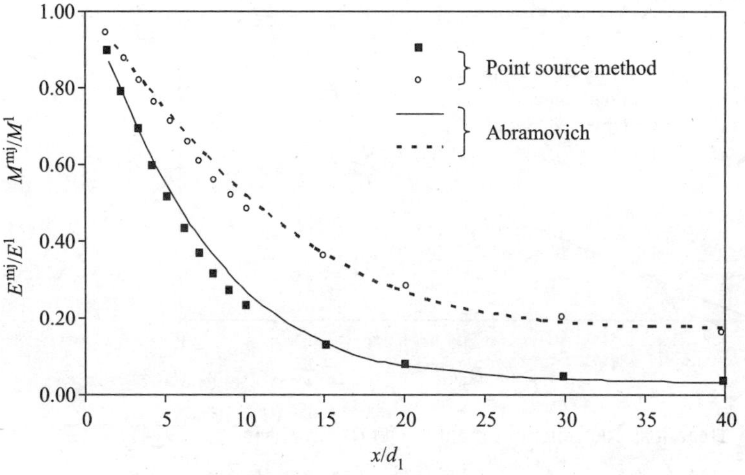

Figure 9.3 Axial variation of the momentum and energy flux in the core of the jet

and energy flux with the distance x in a confined space. Figure 9.3 compares Abramovich's results with the values computed from the point-source method enumerated above. The predictions from the two methods seem to agree well over the entire range of $\frac{x}{d_1}$ shown.

The longitudinal dimension of a manhole, D_m, is usually less than six times the diameter of the inflow pipe. While applying the submerged jet theory to a junction box, the distance x will be replaced by D_m. The energy-loss coefficient K_m predicted for various sizes of the junction box based on the aforementioned two methods is shown in Fig 9.4. Experimental values for the case $d_1 = d_2$ and a step height $S = 2d_1$ are also shown in the figure.

By adding the losses at the entrance to the outlet pipe, the total loss coefficient K was obtained. The variation of K for different sizes of the junction box is shown in Fig 9.5. For sizes up to $\frac{D_m}{d_1} \simeq 4$, the agreement between predicted and the observed values is good.

To find how sensitive the value of K is on the step height S, experiments were conducted with $S = 0$ and $S = 1.33d_1$. Figure 9.6 shows that there is close agreement between the observed values and the ones predicted from the potential-core approach. Further, the empirical expression of Marsalek, $K = \zeta\left(\frac{D_m}{d}\right)$, for $S = 0$, $\frac{D_m}{d_1} \leq 4$ is seen to agree very well with the present results from the submerged jet theory.

Many a time, the diameters of the inflow and outflow pipes of the junction box are not identical. Experiments show that even if the two diameter differ by approximately 25 per cent, the change in the head-loss coefficient K_m is

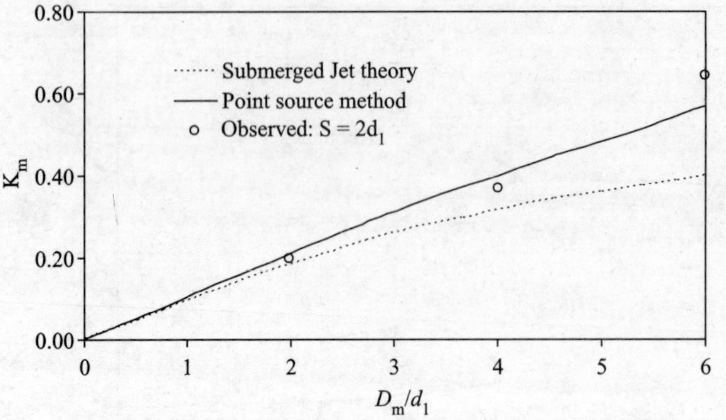

Figure 9.4 Head loss coefficient of manhole jet (Source: Mudgal, 1994)

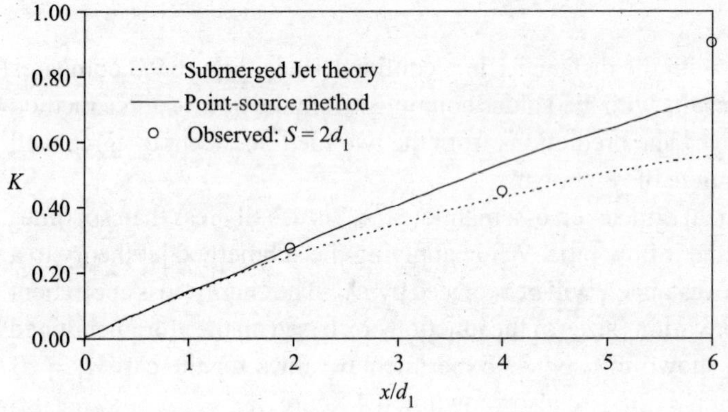

Figure 9.5 Total head loss coefficient K

marginal. In contrast, the head loss coefficient increases significantly if the axes of the pipes have an eccentricity (Mudgal and Pani, 1992). If the eccentricity is of the order of d_1 or more, either in the horizontal or in the vertical plane, the total head-loss coefficient $K \simeq 1.4$.

The depth of submergence, measured from the invert of the inflow pipe, was varied to cover the practical range for the case of $D_m = 4d_1$, $S = 1.33d_1$, and zero eccentricity. The observed values of the loss coefficient K are shown in Fig 9.7 and it can be noticed that when submergence exceeded $2d_1$, the head loss coefficient is independent of the depth of submergence. Similar results have been reported by others (Guymer et al. 2005).

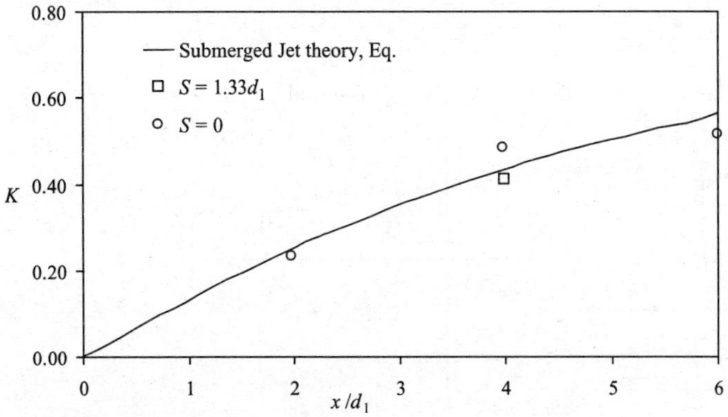

Figure 9.6 Effect of height of step on K (Source: Mudgal, 1994)

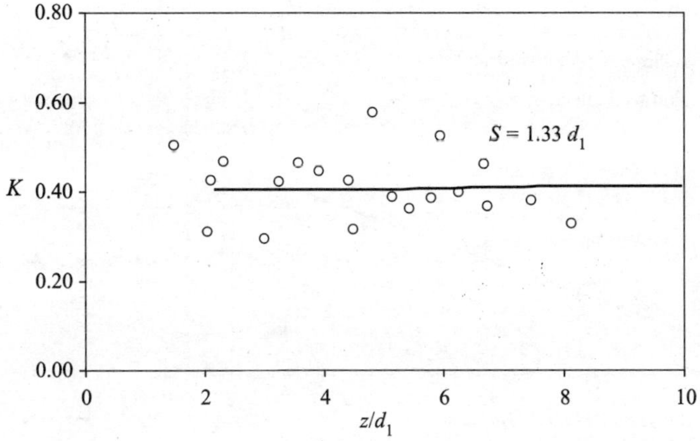

Figure 9.7 Effect of submergence depth on head loss coefficient

While finding the head loss coefficient K in the chamber because of the flow generated by action of the jet, it has been presumed that the end-wall of the box does not interfere with the flow pattern of the jet before it makes an entry into the outflow pipe. To find the extent to which the presence of the end-wall might affect the theoretical results, experiments were repeated for different positions of the entry to the outlet pipe, relative to the end-wall. A schematic representation of the re-entrant type of inlet created by the outflow pipe is shown in Fig 9.8. The step height $S = 2.77d_1$, and $d_1 = d_2$. Further, from Fig 9.9, it can be observed that the proximity of the end-wall plays a very

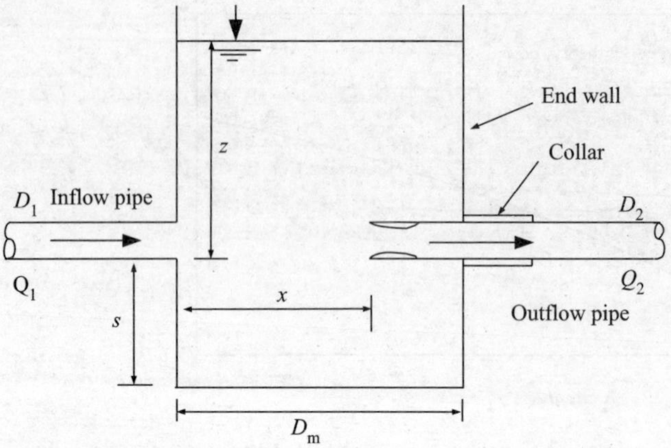

Figure 9.8 Schematic diagram of a re-entrant outlet pipe

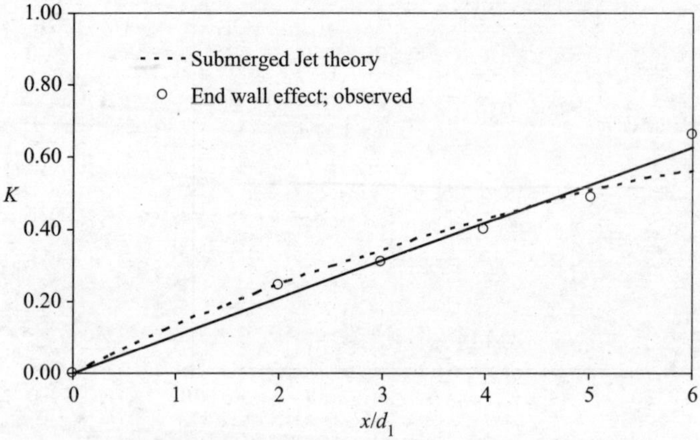

Figure 9.9 Effect of end-wall on coefficient K (Source: Mudgal, 1994)

insignificant role, and the submerged jet theory serves as a good model for the prediction of K.

9.4 Summary

Generally, the length of the junction box is smaller than the length of the potential core of the incoming jet. The pattern of flow in a junction box is quite complex. Yet, it has been established that the coefficient of head loss, K, can be predicted based on the submerged jet theory. The study also reveals that the potential-core method and the point-source method give similar results as regards the loss of energy.

Small differences in the sizes of the inflow and the outflow pipes have insignificant effects on the head-loss coefficient. However, small eccentricities between the axes of the two pipes can increase the value of K significantly.

It seems justified to ignore the presence of the end-wall in the theoretical formulation in which the outlet pipe captures the energy of the core flow in the jet. The value of K is not effected by the depth of submergence if it exceeds two times the pipe diameter d_1.

10 Jet Flocculator

10.1 Introduction

Suspended materials like soil particles, microscopic organisms, colour compounds and organic matter from decaying vegetation and municipal waste etc. are responsible for causing turbidity in water. Water in rivers and reservoirs exhibits wide fluctuation in its quality because of the presence of pollutants and its turbidity. Before supplying water for drinking and other purposes, normally, water treatment is carried out through a series of unit operations – screening, aeration, coagulation, flocculation, sedimentation, filtration and disinfection.

The extent to which the suspended matter absorbs/scatters light, incident on the suspension, is a measure of the turbidity. Turbidity in excess of 5 NTU (Nephelometric turbidity unit) is objectionable because of health and aesthetic reasons. Since the suspended particles are very small, which are often in a colloidal phase, coagulation and flocculation are essential to accelerate their removal. The terms coagulation and flocculation are often used interchangeably though they refer to two distinct processes. Coagulation is the process through which colloidal particles and very fine solid suspensions are destabilised by adding chemicals so that they begin to agglomerate if favorable conditions are created. In contrast, flocculation refers to the process by which the destabilised particles actually conglomerate into larger aggregates and settle in a short duration of time.

The coagulation process, comprising hydrolysis of the coagulant and destabilisation of the colloidal particles, is completed almost instantaneously. Hence, for the coagulation process to be as effective as possible, it is essential that the coagulants be dispersed uniformly in the entire mass of water with great rapidity. This is done by providing rapid mixing units at the head-end of the treatment plant, which generates intense turbulence in the raw water. The rapid mixer can operate based on either mechanical or hydraulic principles. The latter is more suited since no special equipment needs to be installed.

There are several types of hydraulic rapid mixers which are suitable for use in either open channels or in pressure conduits. Open channel mixers are generally simpler to operate and maintain and are cost effective. In the case of pressure conduits, diffusion of jets provides the necessary mechanism of bringing about rapid mixing. The following is a brief description of the arrangement necessary to enhance rapid mixing in the raw-water supply pipeline.

10.2 Rapid Mixing of Coagulants

The intensity and the duration of agitation control the process of rapid mixing. Camp and Stein (1943) introduced a velocity gradient parameter G as a measure of the intensity of agitation in the mixer. The G parameter is defined as

$$G = \sqrt{\frac{P}{\mu \forall}} \tag{10.1}$$

where P is the power dissipated in the water (watts), $\forall$ is the volume of water (m^3) and μ is the absolute viscosity of water $\left(\frac{N-s}{m^2}\right)$. Similarly, the detention time, defined as $T = \frac{\forall}{Q}$, takes care of the duration of agitation. Interestingly, in rapid mixers, the role of G and T seems to be less significant. It is rather important to ensure that the chemical gets thoroughly dispersed quickly and stagnant zones are avoided.

A typical layout to produce agitation hydraulically is shown in Fig 10.1. A baffle plate is kept at a distance of $\frac{1}{3}$ to $\frac{1}{2}$ times the inlet diameter above the mixer floor. The nozzle is located 200 mm above the baffle plate, and the coagulant solution is injected at the point where the nozzle discharges the recirculating water. The pump draws a small portion of the flow and delivers the same through the nozzle. As an alternate method, rapid mixing of the coagulant with the raw water can be achieved by employing an orifice plate placed in the raw-water pipeline. The coagulant solution is fed through a perforated diffuser placed upstream of the orifice plate. Hydraulic mixing is effective if the product GT falls within a range of 300 to 1600 (Kawamura, 2000). If the seasonal flow fluctuations are large for satisfying this criterion, the design should provide two parallel hydraulic mixing systems. Mixing by pressurized water jets seems to be a preferred choice among the alternatives. The possibility of clogging the nozzle by the coagulant and debris present in the water can be a major disadvantage for the mixing device. A typical arrangement for rapid mixing, also known as flash mixing, is shown in Fig 10.2.

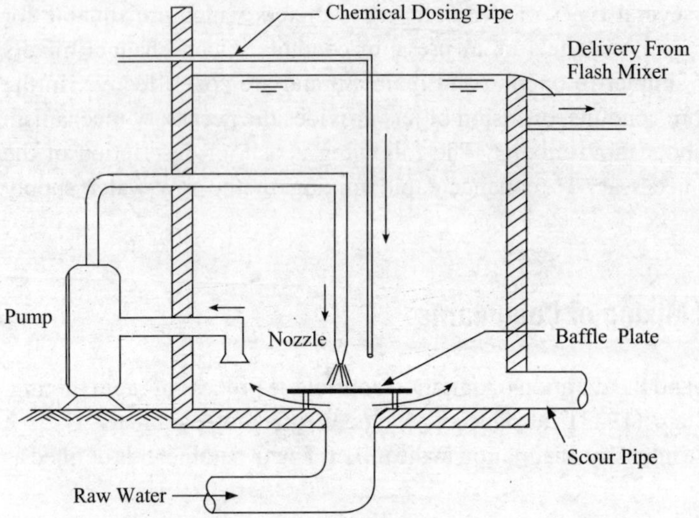

Figure 10.1 Rapid mixing by hydraulic jets

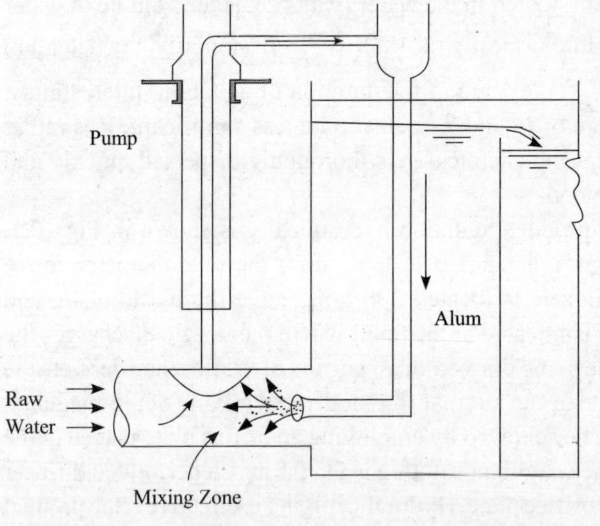

Figure 10.2 Rapid mixing in a pipe by counter flowing jets

10.3 Flocculators

A flocculator is a slow mixing device, and the gentle mixing phase takes place in it. The aim of the operation is to bring about particulate collision and

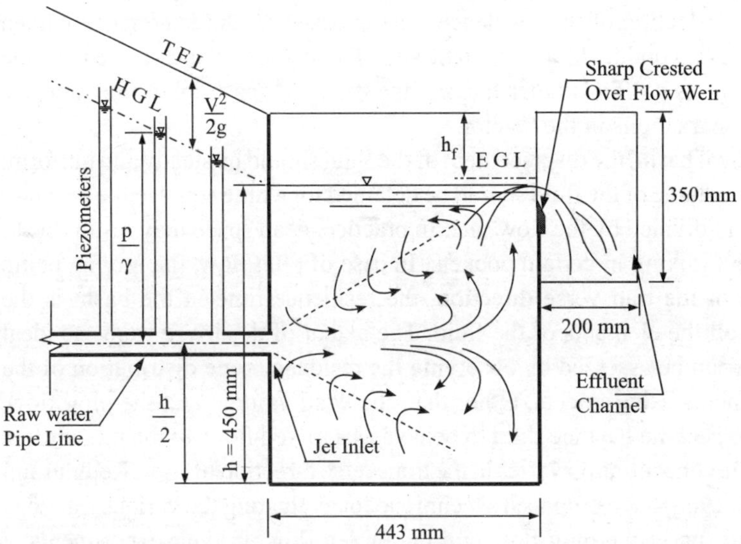

Figure 10.3 Schematic layout of a jet flocculator

enmeshment of particulates into flocs. The choice of a proper flocculation process depends on many factors (Kawamura, 2000). The alternatives for initiating the flocculation process include mixing systems such as mechanical mixing, baffled channel basins etc. However, very little attention has been given to the slow mixing caused by a turbulent water jet, which is fairly efficient. Such an alternative is known as the jet flocculator and the details of its operation are discussed below.

A schematic layout of the jet flocculator is shown in Fig 10.3, wherein the coagulant is supposed to be mixed rapidly in the raw-water supply line itself.

The difference in elevation between the total energy line at the inlet section and the energy line over the overflow weir represents the energy loss in the flocculator. The jet is located at mid-depth and the shape of the tank could be a square, a circle or any other geometric shape. The shape does not seem to have any significant effect on the overall performance.

The residual turbidity of the effluent serves as a simple measure of the degree of flocculation that takes place in the chamber. Though simple to measure, it does not provide information concerning the size of the flocs formed and the dynamics of the formation of aggregates. Alternate methods of measuring the sizes are difficult to adopt. Traditionally, the residence time is considered as an important parameter, while deciding on the size of the flocculation chamber. It indicates the provision of appropriate volume for achieving flocculation.

Often, the selection of the residence time is based on the experience gained from jar tests conducted at the plant site. The choice of the residence time is important as it effects other factors like the dead space, plug flow and the mixed flow fractions in the reactor.

In an ideal basin, the displacement of the fluid should be steady and uniform. Each unit volume of the fluid should be detained for a time equal to the capacity of the basin divided by the flow rate. In practice, dead space may result due to insufficient mixing in certain pockets. In case of plug flow, the mixing being complete in the transverse direction, the residence time in the basin is the same for all the elements of the fluid. The extent of departure from the ideal condition can be assessed by observing the residence time distribution of the fluid elements using tracers. Other than the dead volume and the plug flow, one can expect the balance fluid to behave as a mixed flow. In the mixed flow regime, the concentration varies in the transverse direction of flow. Rebhun and Argaman (1965) have proposed a technique for evaluating the various fractions – the dead space, the plug flow and the mixed flow. In their experiments, a tracer was injected as a slug near the inlet section of the settling chamber and the tracer concentrations in the effluent flow were recorded as a function of time.

In the slow mixer, the liquid needs to be introduced at specific volumetric rates to generate shear that can facilitate the formation of readily settleable flocs. If the shear is too low, flocs may not grow properly. At the same time, if the shear force in the suspension is too large, the flocs will break. Thus, the jet velocity should be varied until the desired result is achieved.

For the layout shown in Fig 10.3, tests were conducted for different values of the retention time ranging from 6 to 20 minutes. All other parameters like the Camp's number GT for rapid mixing in the supply pipeline, the coagulant dosage, influent raw water and its turbidity (100 NTU), the total alkalinity, and the pH values were held constant for all test runs (Patil, 2005). The residual turbidity of the effluent water reflects how efficiently the flocculator performs in removing the suspended particles. From the plot in Fig 10.4, it is observed that the retention time of the flocculator should exceed 10 minutes.

Larger retention time will need a bigger volume of the flocculation chamber and may prove uneconomical. In case of flocculators fitted with mechanical stirrers, the retention time normally varies between 20 to 30 minutes. In addition to being small in size, the performance of the jet flocculator is seen to be relatively insensitive to the efflux velocity of the jet. Hence, in practice, a lot of flexibility exists in the operation of the jet flocculator to meet variable demand patterns. Another interesting observation is that the positioning of the outlet has negligible impact on the performance of the flocculator. Out of the three alternatives shown in Fig 10.5, the outlet provided in scheme 'C' seems to perform marginally better compared to the other two (Patil, 2005).

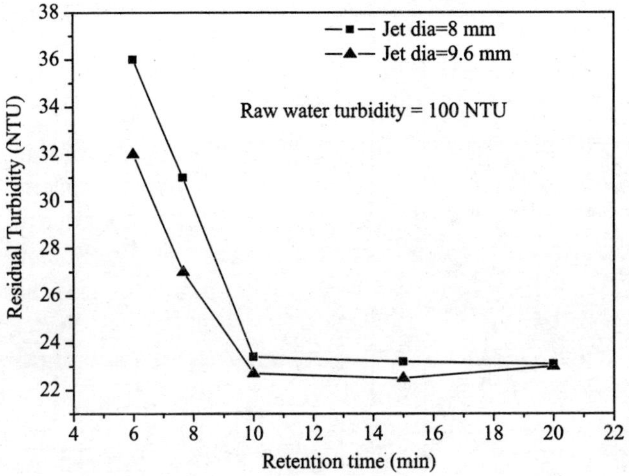

Figure 10.4 Variation of residual turbidity with the basin retention time (Source: Patil, 2005)

Table 10.1 Efficiency of turbidity removal: Retention time = 10 minutes

Basin Type	nozzle dia., mm	Raw water turbidity, NTU	GT	L/d	% removal of turbidity
1A	12.0	100	22600	37	77.7
1B	12.0	100	22800	37	78.6
1C	12.0	100	22600	37	79.4
1C	8.0	100	50680	55.4	78.4

For all the cases tested, the variation in removing the turbidity of the raw water was very small as long as the GT values lay in the range of 22600 to 50600, as indicated in Table 10.1. The range for GT is comparable to the one recommended by Camp (1955) for a mechanically stirred flocculator, that is, 20000 to 60000. The performance drops significantly if the GT value falls below the recommended lower threshold. The turbidity removal of jet flocculators in the laboratory is shown in Table 10.1. The performance is nearly identical to that of many other types of flocculators, including the ones fitted with mechanical stirrers.

If space is not a constraint, the length of the flocculation chamber could exceed $35d$. Experiments show that a minimum length of $35d$ can be selected in the design for economic reasons. As a first approximation, the power dissipated in the chamber can be considered as $P = \rho Q \frac{U_0^2}{2}$. Intuitively, one would expect

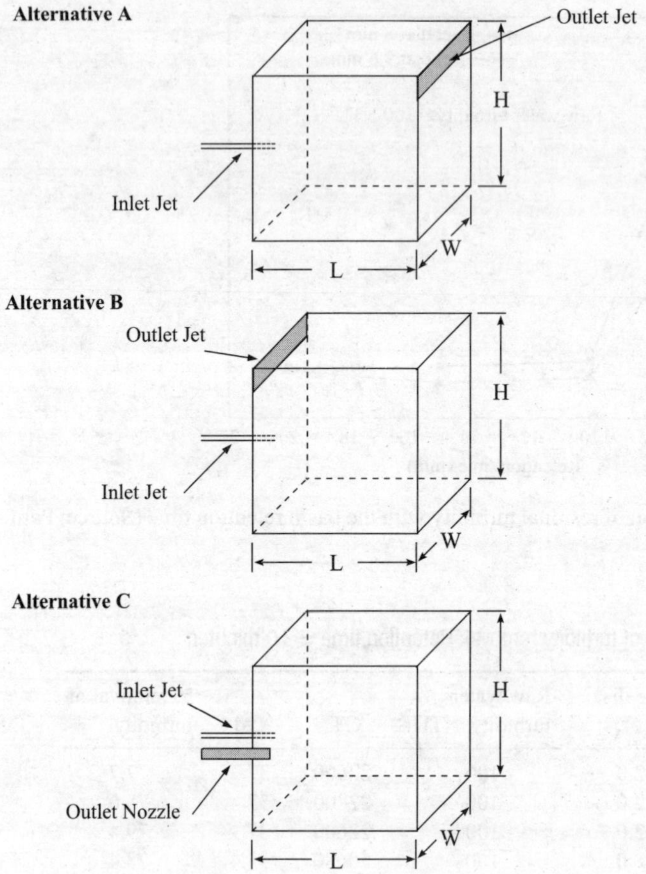

Figure 10.5 Alternate positions of the outlet in a jet flocculator

the plan shape of the basin to have a significant role in assisting turbulent eddies to transfer energy that can affect the formation of flocs. However, in the tests, it was observed that the turbidity removal is nearly identical for all the chamber geometries. Numerical simulation of the turbulent flow in the flocculation chamber, using FLUENT software (Patil, 2005), could offer an explanation for this paradox. The proportion of the dead-space for all the shapes is small and is of the order of 2–3 per cent. The location of the outlet too has very insignificant effects on the dead space. The conventional approach of using the mean flow velocities for finding the energy input to the chamber is not well suited for this analysis. A fraction of the input is only utilized for floc formation. More efficient turbidity removal is associated with larger fractions of utilizable energy.

For comparing the mixing characteristics of a jet flocculator with a flocculator having mechanical stirrers, additional tests were conducted to find the tracer concentration distribution at the exit section. The square tank had a detention time $T = 10$ minutes and the tracer was injected as a slug flow. The shape of the concentration decay curve found by Kawamura (2000), for a tank fitted with a horizontal shaft mechanical stirrer, agrees quite well with the one resulting from the action of jet mixing. Therefore, one could infer that the performances of the two types of flocculators are nearly identical. Actual field performances of static flocculators (Bhole, 1993) and clariflocculators (Armal, 1997) too are nearly identical to those of the laboratory jet flocculator. Hence, the residual turbidity of the jet flocculator scale model can be treated as a fair representation of the value of large size operations. Such a postulate needs further validation in the future. New indices were introduced for explaining the performance of different layouts. However, none of the old, or new indices could offer a complete explanation about the performance of the jet flocculator (Pani and Patil, 2007).

The design procedure to be followed for a single basin jet flocculator can be as follows:

(i) A detention time T is selected in the range of 10 to 20 minutes depending on the expected fluctuation in demand.
(ii) A jet velocity as small as possible is adopted, subject to the constraint $GT \geq 22800$ if Aluminium coagulant is used.
(iii) The jet diameter based on the given flow rate and the jet velocity is calculated.
(iv) The length of the basin is selected as $L = 35d$.

The width W (= height H) is calculated from

$$W = H = \sqrt{\frac{\text{Total volume of the flocculator}}{L}}.$$

(vi) If the height or width of the basin is less than $10d$, the jet may pulsate or adhere to the walls of the chamber due to the Coanda effect. Therefore, a revision in the design, by way of increasing the jet velocity (limited to a maximum of 3 m/s), or increasing the number of basins to two, for parallel operations, is recommended. Normally, this problem may not arise for treatment plants having a capacity of 1 to 2 MLD. The design of a jet flocculator for a small treatment plant will be illustrated by a numerical example.

10.4 Jet Mixing in Storage Tanks

In a water distribution system, storage tanks with large capacities are often used. Long detention times can affect the quality of water adversely. Well mixed storage tanks are desirable for mitigating the adverse effects. Poor locations, orientations and configurations of inlet and outlet pipes can have a detrimental effect. For rainwater harvesting, the rainwater from dwelling roofs is often stored in large tanks for use during the dry seasons, when water is scarce. Water contaminated from roof-top bacteria finds its way into the storage tank. The bacteria count can be reduced through extended storage. Furthermore, for bacteria extinction near the outlet region, mixing of the entire volume of the tank should be avoided. Thus, the requirements of storage tanks are significantly different for the above two purposes. The momentum flux of jets primarily drives the mixing process in the storage tank.

Mixing time of storage tank

For storage tanks in water distribution systems, Rossman and Grayman (1999) have proposed an empirical expression for the mixing time based on experiments. The mixing time is the time needed for the tracer to become essentially homogeneous. For cylindrical laboratory tanks, the mixing time is given as

$$t_m = K\frac{V^{\frac{2}{3}}}{M^{\frac{1}{2}}} \tag{10.2}$$

In Eq 10.2, the constant $K \approx 10.2$, V is the volume of water, and $M = U_0 Q$. Whether the inflow pipe is held vertical or horizontal near the bottom of the tank, the mixing time is not significantly affected. Also, Eq 10.2 describes mixing in full size reservoirs. However, the mixing time t_m is only an overall parameter and provides no detail on the prevailing mixing pattern in the storage tank. With the intent of minimizing the occurrence of dead zones, some researchers have taken recourse to CFD to simulate steady state jet mixing in cylindrical tanks.

From Eq 10.2, one can define a dimensionless mixing time

$$\tau_m = \frac{t_m M^{\frac{1}{2}}}{V^{\frac{2}{3}}} = \text{constant} \tag{10.3}$$

The assumption that τ_m is a constant is only valid for a certain range of $\frac{H}{D}$, where D is the diameter of the tank and H is the water depth. Tian and Roberts (2008) carried out experiments using the 3DLIF techniques to study mixing in different types and shapes of tanks. Using the experimental results, empirical expressions have been proposed for τ_m.

$$\tau_m \approx 10; \qquad \frac{H}{D} \leq 0.8$$

$$\tau_m \approx 10 + 3.5\left(\frac{H}{D} - 1\right); \qquad \frac{H}{D} > 0.8 \tag{10.4}$$

For circular storage tanks, the spatial distribution of the bottom nozzles is important. If a single nozzle is used, it is recommended to place it near the side wall to avoid dead zone formation. A centrally located jet, even in the absence of buoyancy, results in a donut or ring-shaped stable dead zone.

For rectangular tanks with a single nozzle, the horizontal orientation of the jet was most efficient. Choice of location of the nozzle at the centre of the wall or at a corner, is not that crucial. Density difference is temperature dependent. Even a difference as small as 1° C can have a significant impact on mixing. The mixing process can be characterized in terms of the momentum and buoyancy flux. The mixing time can be expressed as

$$t_m = \phi(M, B, D, H) \tag{10.5}$$

where, the buoyancy flux $B = g'_o Q$, and the reduced gravity $g'_o = g\left(\frac{\Delta\rho}{\rho_a}\right)$. In the dimensionless form, the mixing time is written as

$$\tau_m = \frac{t_m M^{\frac{1}{2}}}{D^2} = \phi\left(\frac{H}{D}, \frac{l_M}{H}\right) \tag{10.6}$$

In Eq 10.6, the additional parameter $\frac{l_M}{H}$ accounts for the buoyancy effects. If the water depth $H >> l_M$, the flow will be buoyancy dominated, and the source momentum will play an insignificant role. However, if $H << l_M$, the buoyancy will have very little effect, and the flow will be momentum dominated.

For mixing to take place in the storage tank, the flow should be unstable. The criterion for mixing proposed by Tian and Roberts (2008), is as follows,

$$\frac{dF_d}{H} > 6.4. \tag{10.7}$$

In the above expression, d is the inlet diameter, and F_d is the densimetric jet Froude number. A similar criterion was proposed earlier by Rossman and Grayman (1999), of course with a much smaller value of 1.5 for the constant.

As the volumes of dead zones in the circular and rectangular tanks were small, the mixing results for circular and rectangular storage tanks were generally the same. It was observed that good mixing could be achieved by injecting positively buoyant inflows horizontally at the bottom. In case of negatively buoyant jets, good mixing resulted by directing the inflow horizontally near the water surface. For multiple nozzles, the criterion for mixing can be found by using an equivalent nozzle diameter.

For storage tanks used to store rainwater, water from rooftops may be fed near the bottom of the tank. The user can draw water at the top through a floating outlet. Rainwater enters through the inlet at a lower temperature than the water already in the tank, hence, a jet subject to negative buoyancy will occur. The height, h, at which the the vertical rise of the jet ceases is given as (Fischer et al., 1979)

$$\frac{h}{d} = kF_d \tag{10.8}$$

where k is a coefficient, which lies between 0.5 and 0.7.

In the storage tank, three zones exist–the bottom mixing zone, an aging zone on top of it and the very top freshwater zone. For bacteria decay, it is essential that the height of interface between the mixing and aging zones be minimized. In applying Eq 10.8 to rainwater storage tanks, that part of the original volume of the tank which mixes with the inflow will be treated as h (Martinsen and Lucey, 2004). Experiments were conducted with different jet layouts for minimizing the value of k. For single jets, the lowest value of k was recorded for a jet directed downwards near the tank base. Horizontal jets directed tangentially along the tank wall yielded lower k values than those directed radially from the tank wall. Also, radial manifolds were found to be quite efficient for inhibiting mixing.

10.5 Summary

The single basin jet flocculator is a simple, robust, low cost device, which can be adopted for removal of turbidity from raw water. The retention time is smaller than the one used in mechanically stirred flocculation chambers. The minimum length of the chamber should be approximately 35 times the diameter of the jet. Neither the plan shape of the chamber nor the location of the outlet has any significant effect on the efficiency of operation of the flocculator.

Jets can be used as a mixing device in storage tanks. The location of the inlet and outlet, the orientation of the nozzles and the size of the tank can affect the mixing time. The complex flow pattern in the chamber has been studied by researchers using the 3DLIF technique. These results can be used to validate simulations with available CFD tools.

Design example 10.1:
The capacity of a small treatment plant is specified as 1.0 MLD. The aim is to design a single basin jet flocculator for this treatment plant.

Solution: Capacity $= 1$ MLD $= 41.67\, m^3/hr = 0.01157\, m^3/s$.
Assume $T = 10$ min.
Volume of the basin, $\forall = 0.01157 \times 10 \times 60 = 6.944\, m^3$.
Assume $GT = 30,000$; $\quad G = \frac{30,000}{10\times 60} = 50\ \mathrm{s}^{-1}$
Head loss in the basin is given by $h_f = \frac{G^2\mu\forall}{Q\rho g}$, where $\rho = 997\ \mathrm{kg/m^3}$ (at 25° C), $\mu = 0.89\times 10^{-3}$ kg/m.s, g = 9.81 m/s^s, $Q = 0.01157\ \mathrm{m^3/s}$, $G = 50\ \mathrm{s}^{-1}$, and $\forall = 6.944\ \mathrm{m^3}$
Thus, $h_f = \frac{50^2\times 0.89\times 10^{-3}\times 6.944}{0.01157\times 997\times 9.81} = 0.1365$ m
$U_0 = \sqrt{2gh_f} = \sqrt{2\times 9.81\times 0.1365} = 1.635$ m/s
Cross sectional area of the nozzle, $A = \frac{Q}{U_0} = 0.0070678\ \mathrm{m^2}$
Outlet diameter, $d = \sqrt{\frac{4A}{\pi}} = 0.0948$ m
Provide a jet of diameter $d = 10$ cm.
Check for actual GT value provided:

$$A = \frac{\pi\times 0.1^2}{4} = 0.007854\ \mathrm{m^2} \quad U_0 = \frac{Q}{A} = \frac{0.01157}{0.007854} = 1.473\ \mathrm{m/s}$$

$$h_f = \frac{U_0^2}{2g} = 0.1106\ \mathrm{m}$$

$$G = \sqrt{\frac{Q\rho g h_f}{\mu\forall}} = \sqrt{\frac{0.01157\times 997\times 9.81\times 0.1106}{0.00089\times 6.944}} = 45\ \mathrm{s}^{-1}$$

$GT = 27,000$. Hence, GT value is well within the safe range.

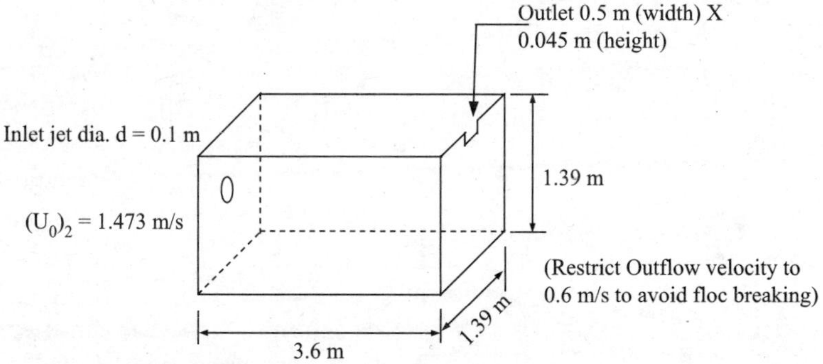

Dimensions of the jet flocculator with a suitable outlet.

Assume a value for the coefficient of discharge of the nozzle as 0.75.
Head required at the jet inlet:

$$\Delta h = \frac{Q^2}{2gA^2C_d^2} = \frac{0.01157^2}{2 \times 9.81 \times \left(\frac{\pi \times 0.1^2}{4}\right)^2 \times 0.75^2} = 0.1966 \text{ m}$$

Select length of the basin, $L = 35 \times d = 35 \times 0.1 = 3.5$ m
Width of chamber = Height of chamber = $\sqrt{\frac{\forall}{L}} = \sqrt{\frac{6.944}{3.5}} = 1.408$ m
Say $W = H = 1.40$ m.
W and H should be larger than $10d$ to avoid an oscillating jet caused by Coanda effect.

To arrive at the actual size of the basin, a free board of 0.15 to 0.20 m should be added to the calculated height H. The schematic layout with dimensions of the tank is shown in the figure.

11 Wall Jets

11.1 Introduction

A wall jet is the shear layer generated when an outlet discharges flow tangentially onto a solid surface. So, the wall jet is bounded by a solid surface (the wall) on one side while the outer region is in contact with the ambient fluid. It has been observed that a little distance away from the wall, free mixing of the fluids takes place. Jets have been classified as curved, cylindrical or radial wall jets depending on the geometry of the wall. Similarly, depending on the shape of the efflux section, two broad categories of jets such as plane wall jets and three dimensional wall jets can occur. All wall jets have been assumed to remain attached to the wall over which they grow.

Wall jets find many applications such as, cooling of surfaces, boundary layer control over flap wings, building ventilation, air conditioning, control of fire, occupational hygiene, etc. Regardless of the opening shape, after a short distance downstream of the outlet, all jets tend to behave in a similar fashion. Prediction of the velocity profiles in a wall jet is more difficult compared to the free jet case, as the effect of molecular viscosity very near the wall can not be ignored. Glauert (1956) treated mathematically the case of a turbulent plane wall jet satisfying the no-slip boundary condition at the wall.

In this chapter, the incompressible plane wall jet will be presented in brief. The three-dimensional wall jet will be discussed in more detail because of its practical utility. The ambient fluid of the wall jet considered in this presentation has the same property as the jet.

11.2 Plane Wall Jet

A schematic diagram of a plane wall jet is given in Fig 11.1. The width of the slot is b_0 and the efflux velocity U_0 is assumed to be distributed uniformly over the nozzle cross-section.

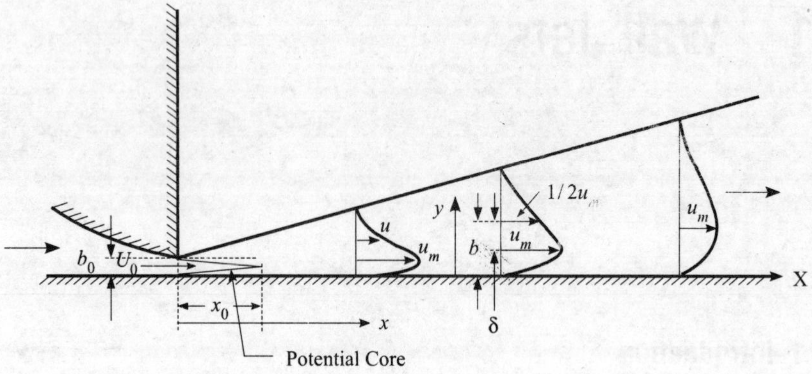

Figure 11.1 Definition sketch of a plane wall jet

A boundary layer which starts from the efflux section develops on the wall and a shear layer develops on the outer side of the nozzle. The flow in the space bounded by the inner edge of the top shear layer and the outer edge of the boundary layer is unaffected by the viscous effects. The maximum velocity is equal to U_0. This wedge shaped fluid element is popularly known as the 'potential core'. If one assumes the shear layer to grow on the inner side at an angle of 5° with the horizontal, the section at which the potential core vanishes under the combined effect of the boundary layer and the shear layer lies at a distance of $\simeq 6.4b_0$.

Distribution of mean velocity

The mean velocity profiles beyond the potential core satisfy similarity when u and y are normalized by a velocity scale 'u_m' and a length scale 'b', respectively. As a convention, b is taken as a value of y where $u = 0.5u_m$ and $\frac{\partial u}{\partial y}$ is negative. The maximum velocity u_m occurs at a distance of $y = \delta$. Historically, Forthman (Rajaratnam, 1976) measured the velocity profiles in a plane wall jet, followed by many others, including the comprehensive work of Bradbury (1965). Verhoff (1963) proposed an empirical equation to describe the similarity profile as

$$\frac{u}{u_m} = 1.48\eta^{\frac{1}{7}}\{1 - erf(0.68\eta)\} \qquad (11.1)$$

where $\eta = \frac{y}{b}$. Figure 11.2 shows the shape of the distribution described by Eq 11.1. It agrees very well with the observed values (not shown here). In the same figure, the Gaussian distribution used for free jets is superposed.

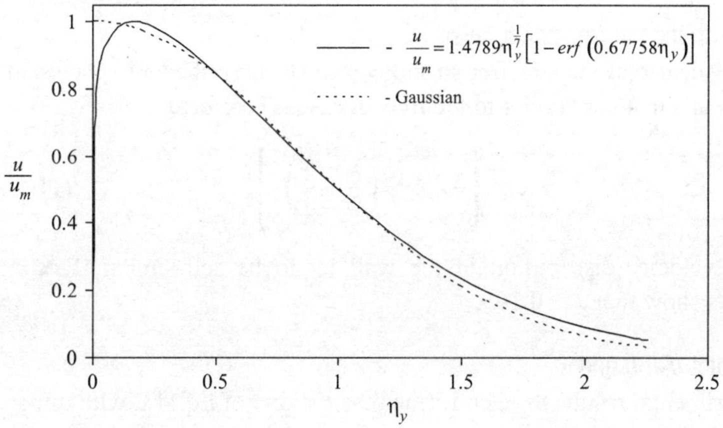

Figure 11.2 Similarity profile: Velocity distribution in a plane wall jet

Interestingly, barring a small region near the wall, the agreement between the free jet and the wall jet velocity distributions is very close.

Exponential relationships

By adopting a procedure identical to the case of a free jet, that is, based on the equations of motion and the integral momentum equation, the following basic laws for the variation of u_m, b and the boundary shear stress τ_0 of a plane wall jet can be derived (Rajaratnam, 1976),

$$u_m \propto \frac{1}{\sqrt{x}}$$
$$b \propto x \tag{11.2}$$
$$\tau_0 \propto \frac{1}{x}$$

Velocity distribution in wall region

Instead of expressing the velocity distribution by a single expression, given in Eq 11.1, the wall jet could be divided into two zones, namely the wall region ($0 < y \leq \delta$) and the outer free mixing region ($\delta \leq y \leq \infty$). The velocity distribution in the wall region of the plane wall jet was given by Myers et al. (1963) as

$$\frac{u}{u_*} = 5.6 log\left(\frac{yu_*}{\nu}\right) + 4.9 \tag{11.3}$$

Equation 11.3 looks very similar to the velocity distribution in turbulent boundary layers.

Velocity distribution in free shear layer

The velocity distribution in the free shear layer can be described by a Gaussian distribution in a manner similar to the free jet case. Therefore,

$$\frac{u}{u_m} = \exp\left\{-0.693\left(\frac{y-\delta}{b-\delta}\right)^2\right\} \tag{11.4}$$

For the velocity distribution in the wall jet to be self-similar, $\delta \propto b$. Experiments show that $\delta \simeq 0.16b$.

How to predict u_m, b and τ_0

Many experimental results have confirmed the validity of Eq 11.2 with minor differences in the coefficients of proportionality. On an average, the variations of the various quantities are as follows:

$$\frac{u_m}{u_0} = \frac{3.5}{\sqrt{\frac{x}{b_0}}} \tag{11.5}$$

and

$$b = 0.068x \tag{11.6}$$

Equation 11.5 is valid for $\frac{x}{b_0}$ extending up to a value of at least 100. Equation 11.6 indicates that the wall jet grows at a rate equal to 70 per cent of the free jet growth.

As regards the variation in the boundary shear stress is concerned, a plot of $\frac{1}{\tau_0}$ versus $\frac{x}{b_0}$, from tests with different Reynolds numbers, results in a linear relationship. However, it will be more useful to express the boundary shear stress in a non-dimensional form, such as

$$c'_f = \frac{\tau_0}{\frac{1}{2}\rho u_m^2} \tag{11.7}$$

Sigalla (1958) recommended an empirical equation,

$$c'_f = \frac{0.0565}{\left(\frac{u_m \delta}{\nu}\right)^{\frac{1}{4}}} \tag{11.8}$$

Adopting the velocity decay relationship of Eq 11.5 in conjunction with the velocity distribution given in Eq 11.1, it was found that

$$\frac{q}{q_0} = 0.248\left(\sqrt{\frac{x}{b_0}}\right) \tag{11.9}$$

where q_0 and q are the discharges per unit width of the nozzle and the unit discharge at the location 'x', downstream of the nozzle, respectively.

An expression for the entrainment velocity, v_e, can be written from Eq 11.9 as

$$v_e = \frac{dq}{dx} = 0.035u_m \tag{11.10}$$

Turbulence characteristics of plane wall jets have been measured by Mathieu and Tailland (Rajaratnam, 1976), and Bradbury (1965).

11.3 Three-dimensional Wall Jets

Flow emanating from different shapes of outlets, like circular, square, elliptical or a bluff rectangular one, results in wall jets which are three dimensional in nature. A schematic sketch of a three-dimensional wall jet is shown in Fig 11.3. The rate of growth of the three dimensional wall jet is significantly different in the y and z-directions. However, the mean velocity profiles in the two directions exhibit similarity and can be expressed as

$$\frac{u}{u_m} = g\left(\frac{z}{b_z}\right) = g(\eta_z) \tag{11.11}$$

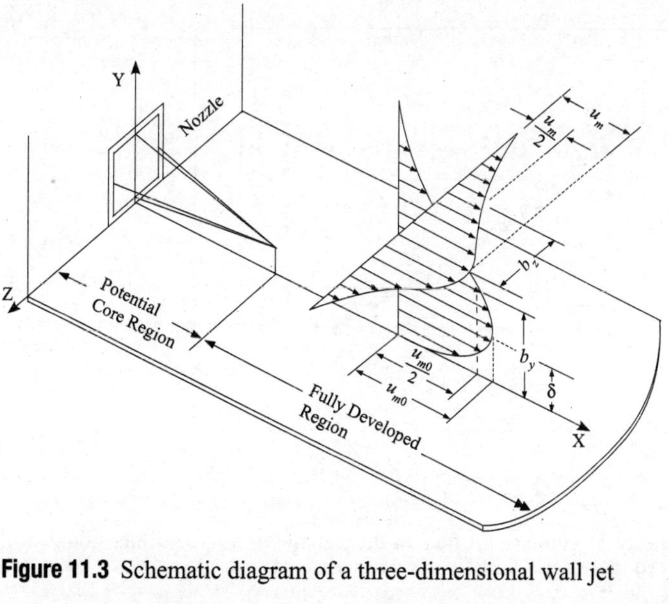

Figure 11.3 Schematic diagram of a three-dimensional wall jet

and

$$\frac{u_m}{u_{m0}} = f\left(\frac{y}{b_y}\right) = f(\eta_y) \tag{11.12}$$

Observed velocity distribution

Typical distribution of the non-dimensional velocity profiles in the z-direction is shown in Fig 11.4. The z-plane is located close to the boundary, that is, at $\frac{y}{d} \simeq 0.16$.

Similarly, the non-dimensional velocity $\frac{u_m}{u_{m0}}$ in the mid-plane, normal to the wall, is plotted against the normalised distance $\eta_y \left(= \frac{y}{b_y}\right)$ in Fig 11.5. Both the above mentioned figures are based on the experimental observation of a wall jet produced by a circular nozzle. Recent measurements of the velocity profiles by Law and Herlina (2002), employing a non-intrusive technique (PIV) are generally in agreement with the earlier observations which relied on the Pitot tube.

In Fig 11.4, the Gaussian distribution for a free jet is also shown and it can be observed that the three-dimensional wall jet diffuses like a free jet in the z-plane, even though spreading takes place at a much faster rate. In contrast, the distribution of u_m in the y-direction, Fig 11.5, exhibits a close agreement between the classical two-dimensional wall jet profile and the experimental values of the three-dimensional wall jet. These observations are empirical in nature and have no theoretical basis.

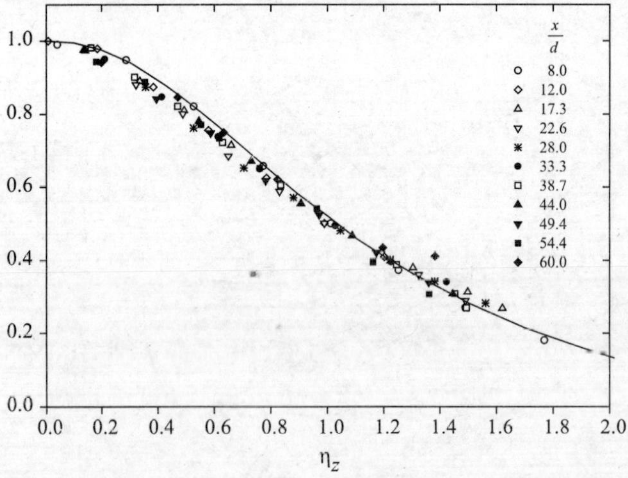

Figure 11.4 Similarity of velocity profiles in the z-plane of a three-dimensional wall jet (Source: Pani, 1972)

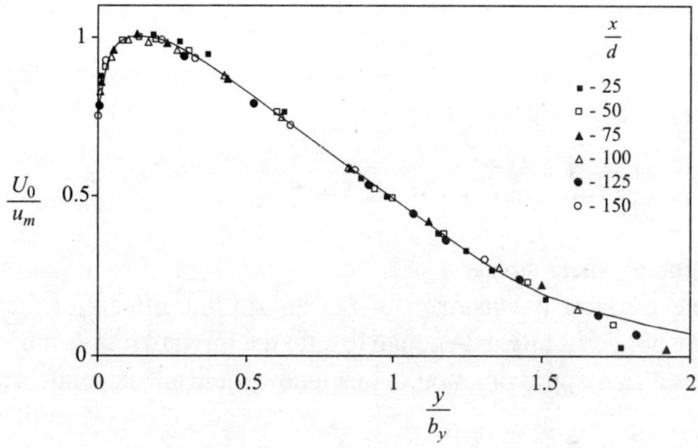

Figure 11.5 Similarity of velocity profiles in the y-plane of a three-dimensional wall jet (Source: Pani, 1972)

The laws that govern the variation of the velocity scale 'u_{m0}' and the two length scales b_y and b_z with x have a theoretical basis. They will be considered in brief as they provide a framework for analyzing/interpreting the available experimental data.

Theoretical analysis

Let us consider the equations of motion for a bluff wall jet making use of the boundary layer type of approximation. The simplified equations are

$$u\frac{\partial u}{\partial x}+v\frac{\partial u}{\partial y}+w\frac{\partial u}{\partial z}=\nu\left(\frac{\partial^2 u}{\partial y^2}+\frac{\partial^2 u}{\partial z^2}\right)-\left(\frac{\partial\overline{u'v'}}{\partial y}+\frac{\partial\overline{u'w'}}{\partial z}\right) \tag{11.13}$$

and

$$\frac{\partial u}{\partial u}+\frac{\partial v}{\partial y}+\frac{\partial w}{\partial z}=0 \tag{11.14}$$

where u, v and w are the mean velocity components in the x, y and z directions, respectively. u', v' and w' are the corresponding fluctuating velocity components. By assuming the Reynolds number $\left(\frac{u_{m0}b_y}{\nu}\right)$ to be large, that is, ignoring the effects of the molecular viscosity, one can express Eq 11.13 in terms of the similarity functions g and f given in Eqs 11.11 and 11.12. On simplification, an equation identical to the three-dimensional free jet (Eq 5.14) will result.

If it is assumed that $u_{m0}\propto x^p$, $b_y\propto x^{q_1}$, and $b_z\propto x^{q_2}$, for the flow to be self-similar, the exponents $q_1=1$ and $q_2=1$ (see section 5.3).

Integral momentum equation

By integrating Eq 11.13 in the limits $z = -\infty$ to ∞, and $y = 0$ to ∞, the integral momentum equation works out as

$$\frac{d}{dx}\left(\int_0^{\infty}\int_{-\infty}^{\infty}\rho u^2 dydz\right) = -\tau_0 \tag{11.15}$$

where τ_0 is the boundary shear stress.

For a reasonable distance downstream of the nozzle, the effects of the boundary shear may not be significant enough to alter the forward momentum of the jet. Hence, as a first approximation, let us ignore the wall shear stress. Consequently,

$$\frac{d}{dx}\left(\int_0^{\infty}\int_{-\infty}^{\infty}\rho u^2 dydz\right) = 0 \tag{11.16}$$

From Eq 11.16 and the values of q_1 and q_2 already found, $p = -1$. Note that these exponents are identical to the ones found for three-dimensional free jets. This is expected as we have ignored the effects of the wall shear stress.

Dimensional analysis

Dimensional analysis will be useful to find the boundary shear which had been ignored so far in the governing equation. Let us write

$$\tau_{0m} = \phi(M_0, \rho, x, \mu) \tag{11.17}$$

where M_0 is the specific momentum flux of the nozzle and equals AU_0^2. The two non-dimensional parameters that can be framed from the above set of variables are written as

$$\frac{\tau_{0m}}{\frac{1}{2}\rho U_0^2}\left(\frac{x}{\sqrt{A}}\right)^2 = \phi\left(\frac{U_0\sqrt{A}}{\frac{\mu}{\rho}}\right) \tag{11.18}$$

Thus, for a given efflux condition, $\tau_{0m} \propto \frac{1}{x^2}$. Further, if the Reynolds number of the jet is high, the coefficient of skin friction c_f is a unique function of $\frac{x}{\sqrt{A}}$. Note that the choice of $\sqrt{A}$ as a characteristic dimension of the outlet eliminates the effect of the shape of the nozzle.

Skin friction coefficient

Since $u_{m0} \propto x^{-1}$ and $\tau_{0m} \propto \frac{1}{x^2}$, a local skin friction coefficient

$$c_f' = \frac{\tau_{0m}}{\frac{1}{2}\rho u_{m0}^2} \tag{11.19}$$

will attain a constant value in the fully developed zone. The variations of c_f and c'_f with $\left(\frac{x}{\sqrt{A}}\right)$ are shown in Figs 11.6 and 11.7.

Experimental observations making use of the Preston tube technique show that $c'_f \simeq 0.0065$. In contrast, c'_f values computed form the wall function (Law and Herlina, 2002) yield a much higher value of approximately 0.01.

Lateral distribution of boundary shear stress

In the fully developed zone, that is, beyond $\frac{x}{d} \simeq 12$, the distribution of the boundary shear τ_0 in the transverse direction satisfies similarity. Figure 11.8 shows the boundary shear stress distribution measured by the Preston tube technique. The normalised distance $\eta_\tau = \frac{z}{b_\tau}$, where b_τ represents that value of z at which $\tau_0 = \frac{\tau_{0m}}{2}$. For a circular outlet, b_τ is a linear function of the distance x. From measurements, it was found that $\frac{b_\tau}{d} = 0.093 \times \frac{x}{d}$. Hence, in Fig 11.8 the normalised distance is expressed as $\frac{z}{x}$.

Loss of forward momentum flux

Based on the observed boundary shear distribution and the linear growth of the length scale b_τ with x, the total loss in the forward momentum flux was evaluated in a distance of $\frac{x}{d} = 0$ to 100. The results show that omission of the wall shear stress effects results in an error of about 5 per cent in the predicted value of u_{m0} at the location $\frac{x}{d} = 100$. It means that the forward momentum flux drops to approximately 90 per cent of the source momentum flux. Although this estimate is significantly less than those of other researches, this observation helped to ignore the boundary shear in the near field region. Thus, one can develop expressions for the decay of the maximum velocity of three-dimensional wall jets as an extension to the free jet results. Prior to it, however, let us see how the experimental observations describe the decay of the maximum velocity and the growth rate of the shear layers in the y and z directions.

Observed growth rates in the *y* and *z* directions

The variations of b_y and b_z with the streamwise distance x are shown in Figs 11.9 and 11.10. These can be represented by the following relationships:

$$\frac{b_y}{d} = 0.90 + 0.045\left(\frac{x}{d}\right) \tag{11.20}$$

and

$$\frac{b_z}{d} = 0.20\left(\frac{x}{d}\right) - 1.25 \tag{11.21}$$

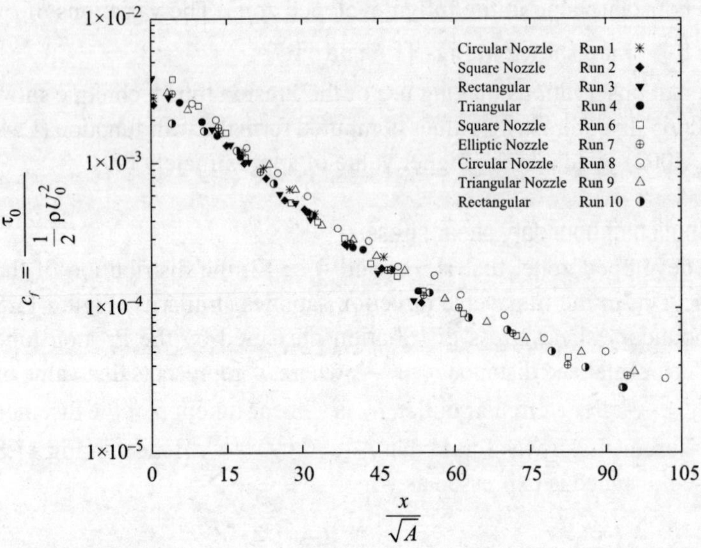

Figure 11.6 Variation of the coefficient of skin friction c_f with distance x (Source: Pani, 1972)

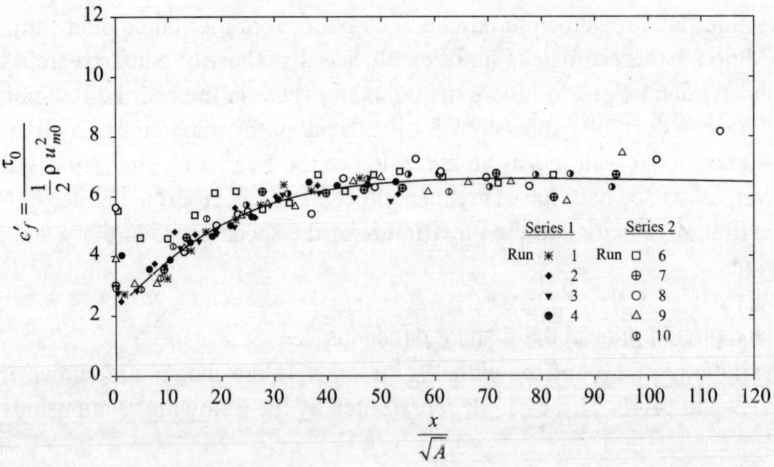

Figure 11.7 Variarion of the local skin friction coefficient c'_f (Source: Rajaratnam and Pani, 1974)

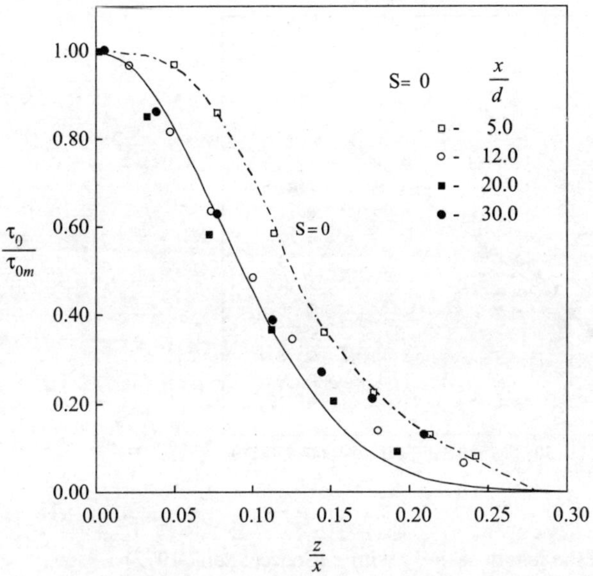

Figure 11.8 Distribution of the boundary shear stress in the $y = 0$ plane (Source: Pani, 1972)

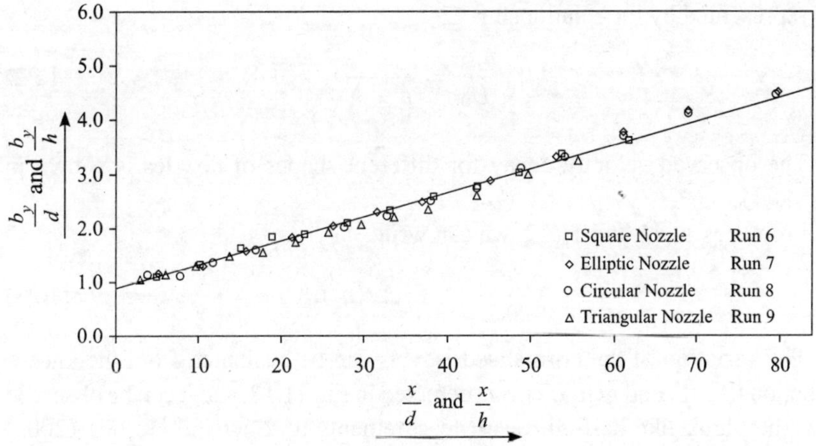

Figure 11.9 Variation of the length scale b_y with x (Source: Pani, 1972)

For rectangular outlets, the height of the outlet b will replace d in Eqs 11.20 and 11.21. The coefficients in the two equations can vary marginally depending on the nozzle details, the measuring technique and the experimental conditions.

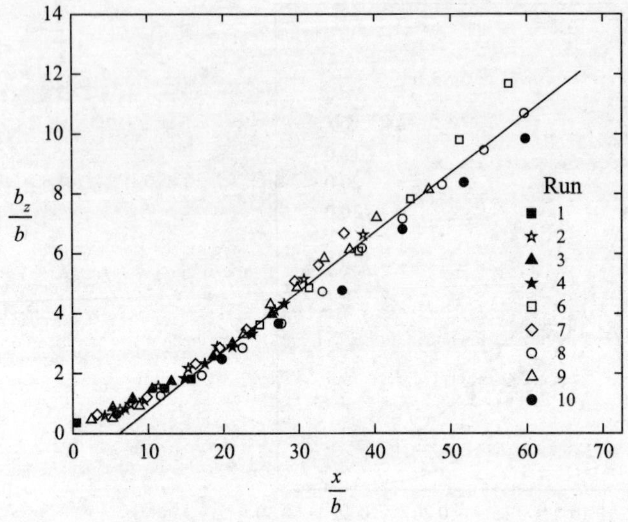

Figure 11.10 Variation of the length scale b_z with x (Source: Pani, 1972)

Decay of the maximum velocity u_{m0}

The maximum velocity decays linearly with the streamwise distance x. It can be represented by the relationship

$$\frac{u_{m0}}{U_0} = \frac{c}{\left(\frac{x}{\sqrt{A}}\right)} \tag{11.22}$$

The observed velocity decay for different shapes of nozzles is shown in Fig 11.11.

From Eqs.11.11 and 11.12, we can write

$$\frac{u}{u_{mo}} = f(\eta_y)g(\eta_z). \tag{11.23}$$

The variation of the normalised isovels can be evaluated from the known functions $f(\eta_y)$ and $g(\eta_z)$. This is depicted in Fig 11.12, and it can be observed that they look like half-ellipses. Measurements by Law and Herlina (2002) using non-intrusive techniques reaffirmed the said observations.

11.4 Application of Reichardt's Hypothesis

The shear resistance offered by a smooth wall is relatively small. Therefore, it looks reasonable to ignore its effects on the loss of the forward momentum

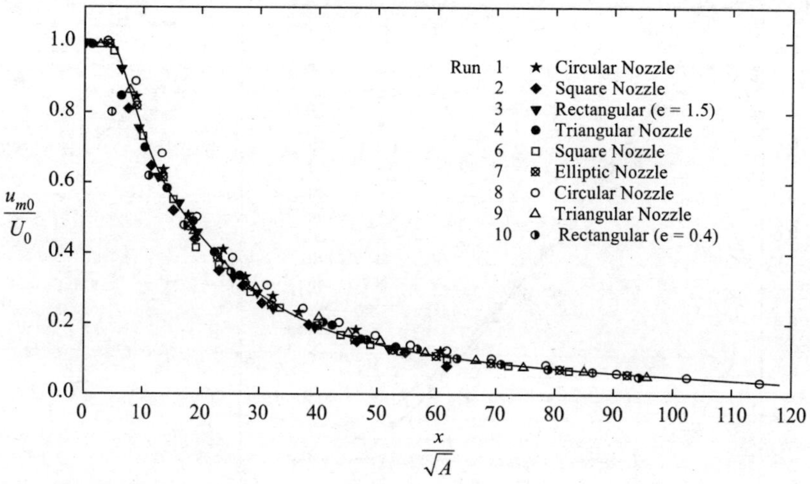

Figure 11.11 Decay of the maximum velocity for different shapes of nozzles (Source: Rajaratnam and Pani, 1974)

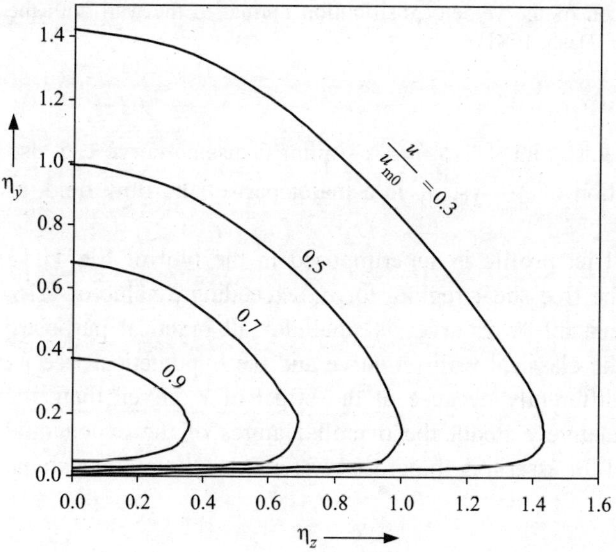

Figure 11.12 Normalised isovels downstream of a circular outlet (Source: Pani, 1972)

flux of a three-dimensional wall jet. Especially for $\frac{x}{d} \leq 100$, this assumption seems reasonable. For flow from a bluff jet, the velocity distributions in the x–y mid-plane and the x–z plane at a height of $y = \delta$, exhibit interesting features. The Gaussian curve, $\frac{u}{u_m} = e^{-0.693\eta_z^2}$, seems to be a good fit to the

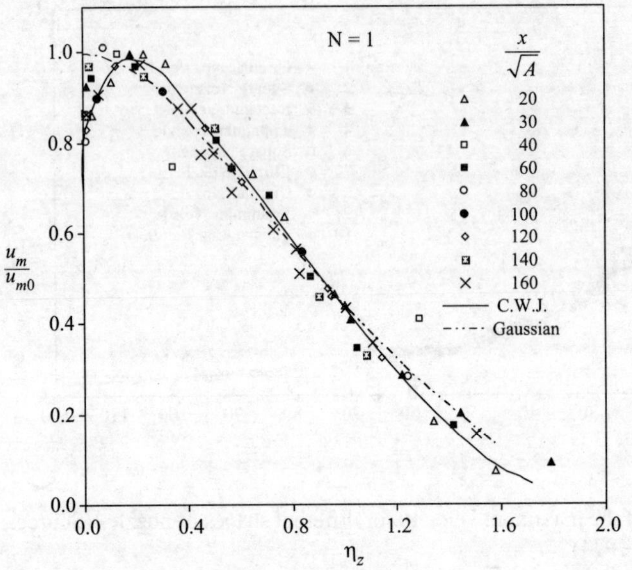

Figure 11.13 Comparison of the velocity distribution normal to the wall with the Gaussian curve (Source: Dash, 1981)

non-dimensional velocity plot $\frac{u}{u_m}$ vs. η_z. A similar Gaussian curve can also describe the distribution of $\frac{u_m}{u_{m0}}$ vs. η_y in a major part of the flow field, as depicted in Fig 11.13.

The classical wall jet profile is superimposed in the plot of Fig 11.13 for comparison. In the free shear region, for η_y exceeding a value of 0.16, the difference between the two curves is small for all practical purposes. In the wall region, the classical wall jet curve and the hypothetical free jet curve behave very differently because of the effect of τ_0. Even then, the wall region being relatively small, the overall features of the exponential curve can be adopted for assessing the gross behaviour of three-dimensional wall jets.

Wall as a smooth reflector

As a smooth wall acts like a reflector, the hypothetical maximum velocity on the wall (in reality the maximum will occur at a small distance δ) can be evaluated based on Reichardt's hypothesis. A procedure similar to the case of the three-dimensional free jet can be followed. Equation 5.58 can be adopted for the case of the wall jet by merely substituting $2A$ in place of A. Therefore,

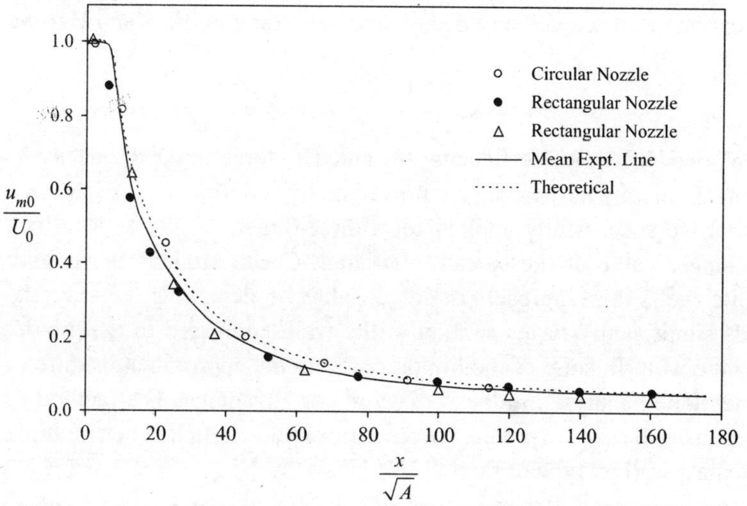

Figure 11.14 Decay of maximum velocity for different shapes of three-dimensional wall jets

the maximum velocity decay relationship for a three-dimensional wall jet is

$$\frac{u_{m0}}{U_0} = \frac{0.94}{C}\left\{\frac{x}{\sqrt{A}}\right\}^{-1} \tag{11.24}$$

In the above expression $C = \sqrt{C_1 C_2}$, where C_1 and C_2 are the coefficients in the half-width relationship $b_y = C_1 x$ and $b_z = C_2 x$. In a wall jet, these coefficients attain constant values at some distance away from the nozzle. The coefficient C_1 attains a near constant value of 0.06 rather quickly, whereas C_2 attains a value of 0.25 asymptotically with the distance x. Overall, the variation in the value of C is relatively insignificant for $\frac{x}{\sqrt{A}} > 50$. The decay of the maximum velocity u_{m0} for the three dimensional wall jet, computed from Eq 11.24, is shown as a dotted line in Fig 11.14.

Effect of shape of outlet

The mean curve obtained from many more experimental observations on three-dimensional wall jets, emanating from different shapes of outlets, is also shown in Fig 11.14. The close agreement between observed data from different sources and Eq 11.24 amply justifies the efficacy of treating the wall as a reflector. In place of an exponent −1.0 in Eq 11.24, some researchers have used a slightly different exponent in the velocity decay law. For example, Law and Herlina proposed $\frac{u_{m0}}{u_0} = 9.23\left(\frac{x}{d}\right)^{-1.066}$

Spread coefficient C

The use of two different coefficients, C_1 and C_2, for finding the decay of the maximum velocity u_{m0} can be circumvented by adopting the *point-source concept* that we successfully applied for a three-dimensional free jet. Even though a single value of the spread coefficient C can predict the velocity decay quite well, this approach is not capable of depicting the way the jet spreads significantly faster parallel to the wall, compared to the growth normal to the wall. In spite of the limitations, and the approximate nature of the forementioned analysis, it does offer a major advantage. The method of superposition can be used to predict the mean flow velocity field when multiple outlets create a wall jet type of flow.

11.5 Wall Jets from Multiple Outlets

It has been observed that the mean velocity field downstream of multiple outlets is quite similar and does not depend on the shape of the nozzles in the array.

Method of superposition

The maximum momentum flux on a frictionless wall, resulting from N outlets, can be expressed by the method of superposition as

$$\frac{u_{m0}^2}{U_0^2} = \frac{0.884A}{b_y b_z} \sum_{i=-J}^{i=J} \exp\left\{-1.386\left(\frac{si}{b_z}\right)^2\right\} \tag{11.25}$$

where b_y and b_z are the half-velocity length scales and $J = \frac{N}{2}$. This relationship, Eq 11.25, can be integrated for infinitely large values of N in terms of an equivalent slot width b_e. The final form of the equation looks as follows:

$$\frac{u_{m0}^2}{U_0^2} = \frac{1.325 b_e}{C_1 x} \tag{11.26}$$

Note that this expression looks similar to the case of a plane free jet, $\frac{u_m}{U_0} = \frac{C}{\sqrt{\frac{x}{b_0}}}$, where b_0 is the width of the slot. However, the constants in the two cases are quite different.

The two length scales, b_y and b_z, found from experiments with a single rectangular wall jet, have been used in Eq 11.25 to evaluate the variation of $\frac{U_0^2}{u_{m0}^2}$

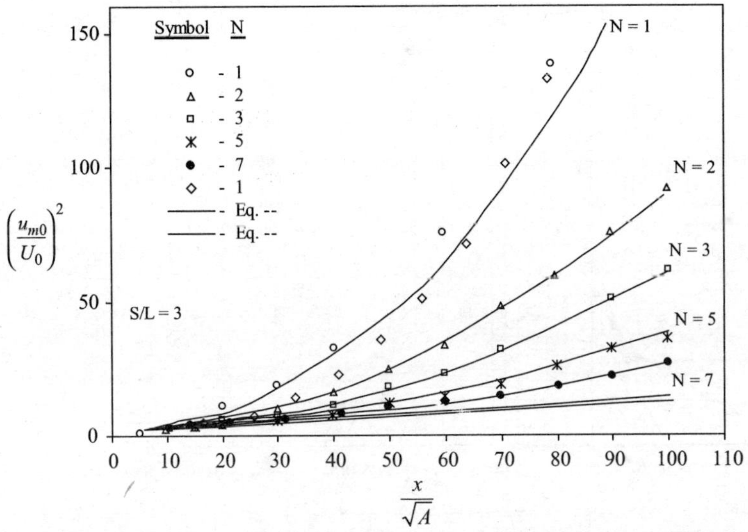

Figure 11.15 Variation of the dynamic pressure for multiple rectangular outlets (Source: Pani and Dash, 1983)

with the distance $\frac{x}{\sqrt{A}}$. Furthermore, for different N in the array, a fair agreement between the computed and the observed values is evident from Fig 11.15. As N becomes very large, the linear relationship described in Eq 11.26 is satisfied.

Observed velocities for $N = 3$

As a typical case, the distribution of the mean velocity in a z-plane for $N = 3$ is shown in Fig 11.16. It is seen that the predicted curves and the observed distribution agree quite well, especially for $x \geq 15\sqrt{A}$.

Similarly, the velocity distribution along the y-axis is shown in Fig 11.17. As in the case of a single wall jet, the classical wall jet curve describes closely the velocity distribution of multiple wall jets. Since the analysis of the wall jet is akin to the free jet, the length scale $\sqrt{NA}$ can still be used to describe the decay of the maximum velocity in a unified way. This is evident from the plot given in Fig 11.18 for different values of N.

11.5.1 Turbulence Characteristics of a Three-dimensional Wall Jet

The fluctuating velocities were measured by Fujisawa and Shirai (1989) using a HWA. The Reynolds number in the tests was in the range 2×10^4 to 6×10^4 and a square shaped nozzle was used to create the jet. The three components of the

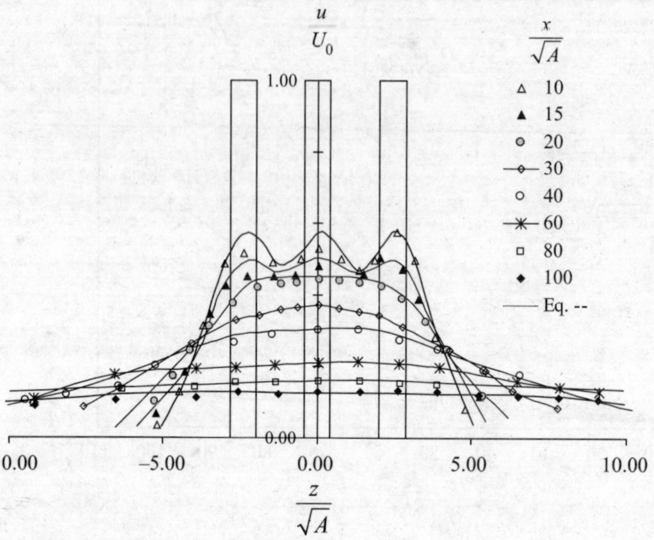

Figure 11.16 Mean velocity distribution in the transverse direction for N = 3 (Source: Dash, 1981)

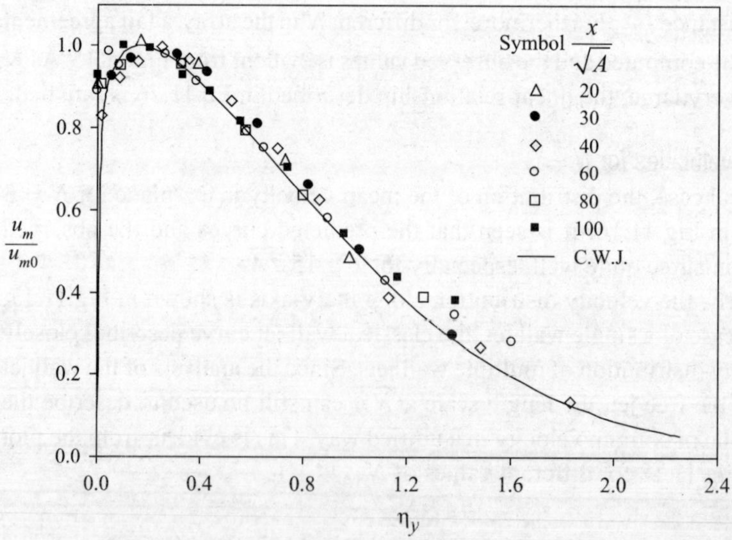

Figure 11.17 Mean velocity distribution normal to the wall for N = 3

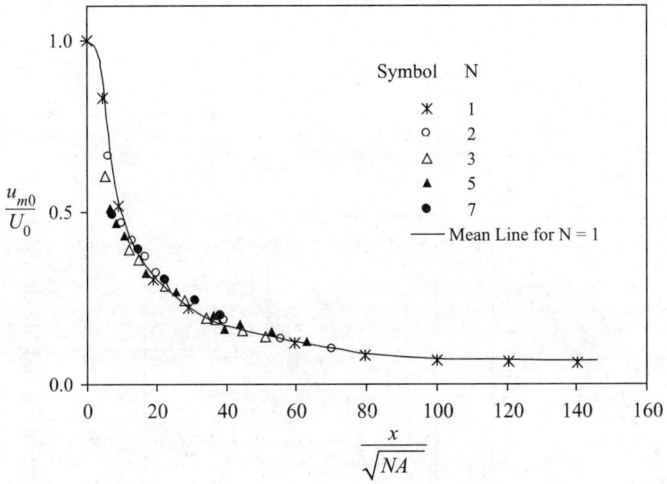

Figure 11.18 Unified maximum velocity decay plot for multiple jets (Source: Pani and Dash, 1983)

turbulence intensities and the Reynolds stress were measured at selected values of $\frac{x}{d}$ in the asymptotic region of the three-dimensional jet. The fluctuating quantities were observed to be in a state of self-preservation. The intensities of turbulence lie between the values of the two-dimensional wall jet and the axially symmetric free jet. In the outer region of a three-dimensional wall jet, the distribution of turbulence over the depth is very nearly the same as that of a free jet. On the contrary, in the inner region of the shear layer, the distribution of turbulence intensity is close to that of a plane wall jet. Due to damping, the turbulence fluctuations $\overline{v'^2}$ become exceedingly small near the wall. Hence, in the wall region, there is anisotropy in the magnitude of the fluctuations as seen in Fig 11.19. One can observe that $\overline{u'^2} > \overline{w'^2} > \overline{v'^2}$.

The distribution of the turbulence intensities and the Reynolds stress in the transverse direction near the wall is shown in Fig 11.20. The wall damping effect not only brings down the magnitude of the fluctuations for all the z-positions, but also creates anisotropy. It can be seen that in the off-centre region, $\overline{w'^2} > \overline{u'^2} > \overline{v'^2}$. It explains the reason for having a large spread of the jet in the transverse direction.

The Reynolds stress in the transverse plane is even smaller than the one in the normal plane. Not very far from the outlet, the mean velocity components attain self-similarity. In contrast, the fluctuating components attain self similarity at a much larger downstream distance from the efflux section.

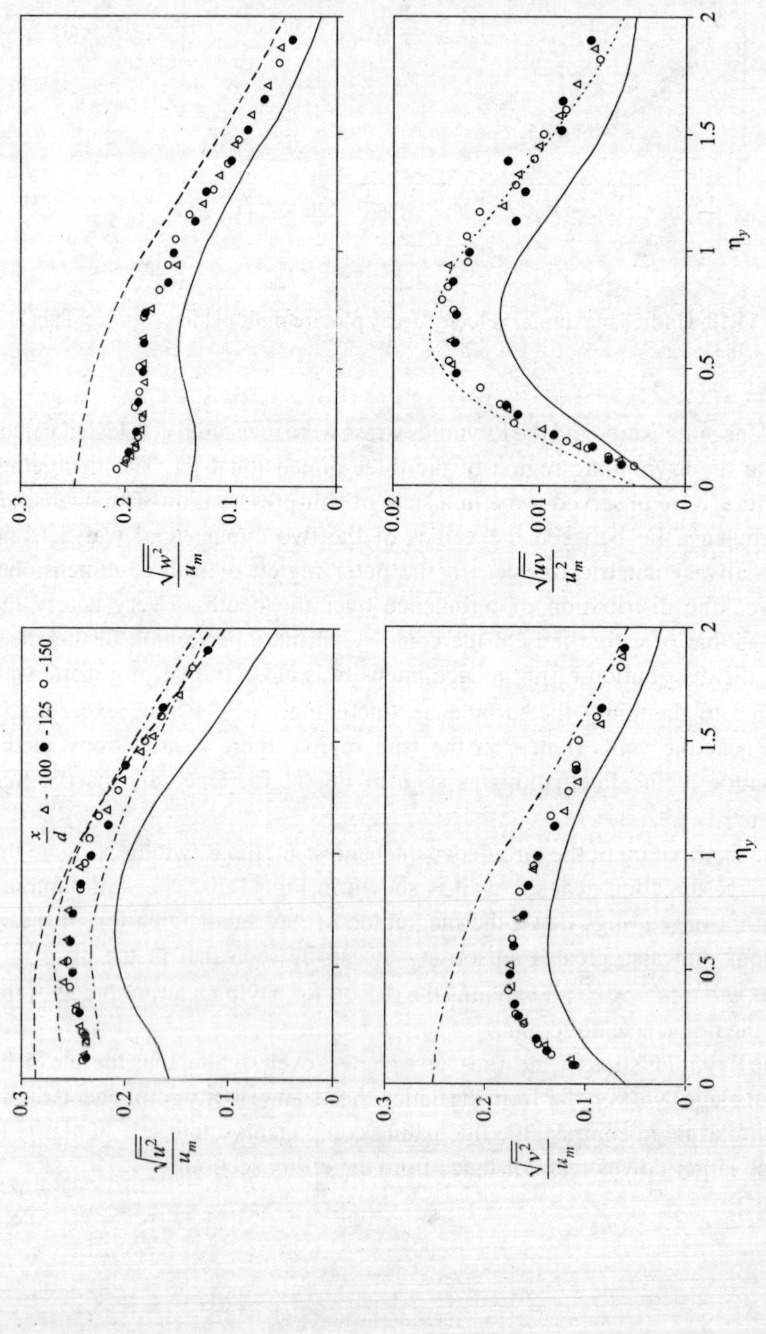

Figure 11.19 Distribution of turbulence and Reynolds stress normal to the wall (Source: Fujisawa and Shiraj, 1989)

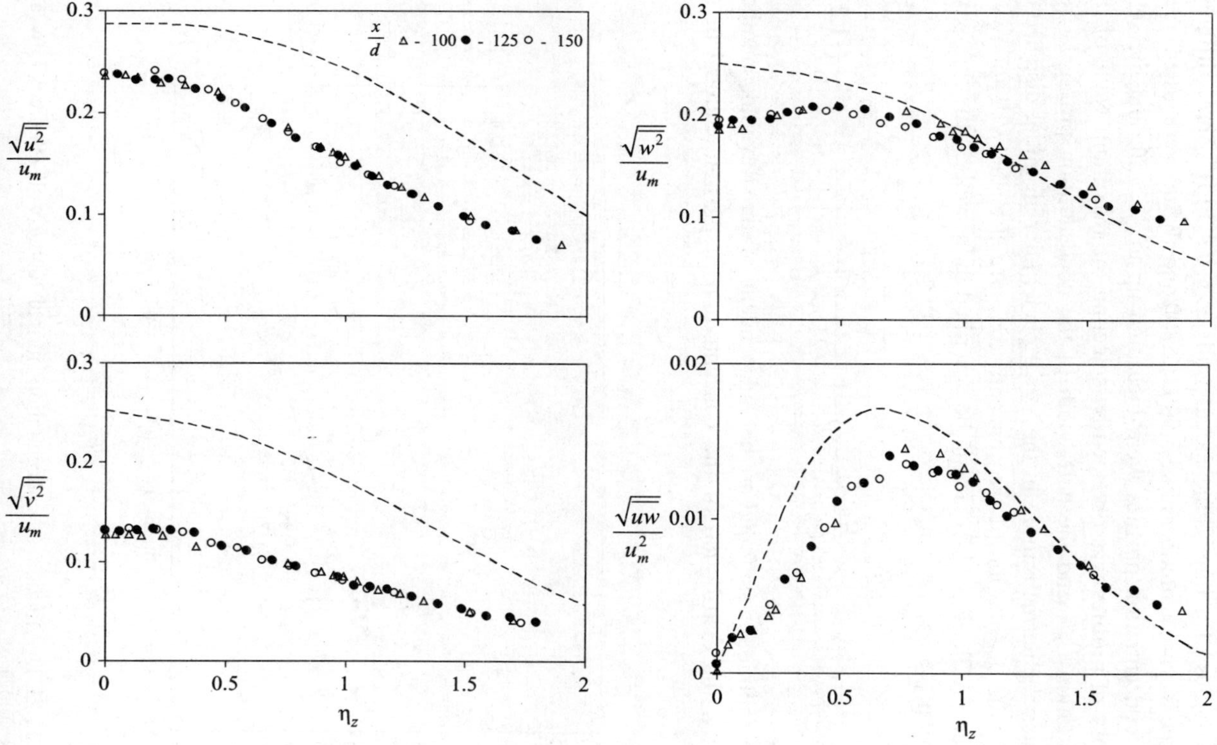

Figure 11.20 Distribution of turbulence and Reynolds stress parallel to the wall (Source: Fujisawa and Shiraj, 1989)

11.6 Concentration Distribution and Decay of Maximum Concentration

Law and Herlina (2002) conducted experiments on the concentration characteristics of a three-dimensional wall jet using PLIF. Three different discharge Reynolds numbers like 5,500, 12,200 and 13,700 were tested.

The concentration profiles were measured up to $\frac{x}{d} = 50$, and the self-similarity of profiles was obtained by selecting the maximum value of the concentration C_{m0} and the half width of the concentration profile b_{yc} as the scales for normalisation. In the $z = 0$ plane, after an initial distance of $\frac{x}{d} = 20$, the distribution of the concentration normal to the wall follows a Gaussian profile, Fig 11.21. It can be described quite well by the following expression,

$$\frac{C_m}{C_{m0}} = \exp - \left(\frac{y}{1.21 b_{yc}}\right)^2 \tag{11.27}$$

The distribution was found to be independent of the Reynolds number. Similarly, the distribution of the concentration in the z-direction can be

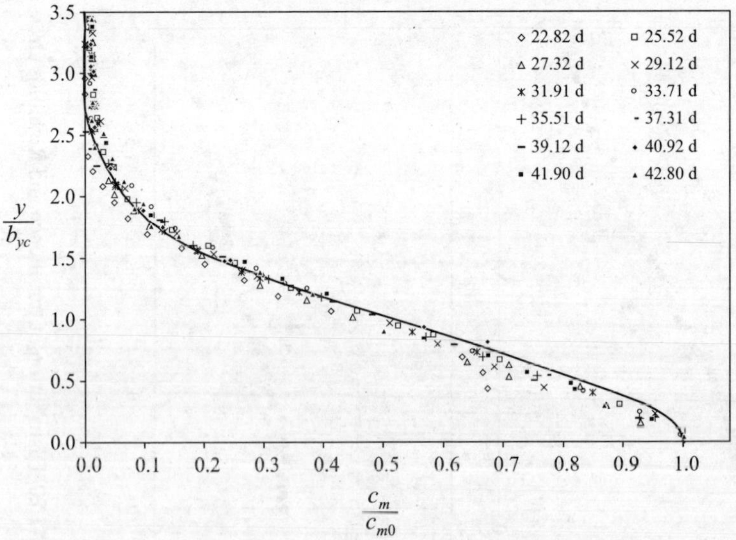

Figure 11.21 Distribution of concentration normal to the wall (Source: Law and Herlina, 2002)

described by the equation

$$\frac{C}{C_m} = \frac{1}{\left(0.965\frac{z}{b_{zc}}\right)^2 + 1} \tag{11.28}$$

Interestingly, the shape of the profile is identical to its velocity counterpart.

Decay of maximum mean concentration

Based on the experimental observations, the best-fit curve for the decay of the maximum mean concentration C_{m0}, beyond $\frac{x}{d} = 20$, is given as

$$\frac{C_{m0}}{C_0} = 6.97\left(\frac{x}{d}\right)^{-1} \tag{11.29}$$

The variation of the concentration length scales are given by the following linear relationships,

$$\frac{b_{yc}}{d} = 0.061\frac{x}{d} + 0.31 \tag{11.30}$$

and

$$\frac{b_{zc}}{d} = 0.31\frac{x}{d} - 3 \tag{11.31}$$

From the above two equations it can be observed that the ratio of the concentration length scales in the two directions is equal to 5.0. This is identical to the ratio of the velocity length scales in the two directions. However, the tracer concentration expands at a much faster rate. Hence, the decay of the maximum concentration C_{m0} is faster compared to the maximum velocity decay. It may be noted that the locations of the maximum values of the velocity and concentration in their respective distributions are quite different. The maximum concentration occurs on the wall, whereas the maximum velocity is located at a distance of $\delta = 0.16b_y$ from the wall.

11.7 Three-dimensional Wall Jets in Coflowing Streams

At times, effluents are discharged as multiple coflowing wall jets. To understand the flow features of such an arrangement, it is essential to understand the behaviour of a single three-dimensional jet in a coflowing stream. Since the shape of the bluff jet is not of great importance, we shall discuss the results pertaining to a circular wall jet investigated by Rajaratnam and Stalker (1982). Compared to circular wall jets in a stagnant ambient, the coflowing velocity U_a significantly inhibits the growth of the jet in the transverse direction.

Experimental observations

The experimental study covered a wide range of velocity ratios, $\beta = \frac{1}{30}$ to $\frac{1}{2}$, and the depth of flow in the channel was varied from 10 to 30 times the nozzle diameter. Being a three-dimensional wall jet, two length scales b_y and b_z were adopted. Here, the definition of the length scale b_y corresponds to the outer free shear layer, i.e. $b_y = (b_{0.5} - \delta)$. Using these half-velocity scales, the velocity profiles of all the runs were described by the standard exponential relationship

$$\frac{\Delta u_m}{\Delta u_{m0}} = e^{-0.693\eta_y^2} \tag{11.32}$$

and

$$\frac{\Delta u}{\Delta u_m} = e^{-0.693\eta_z^2} \tag{11.33}$$

The agreement between Eqs 11.32 and 11.33 and the observed values is good, as shown in Fig 11.22.

A suitably defined momentum thickness θ may be used to describe uniquely the variation of the length scales b_y and b_z. Apart from some scatter in the experimental data, the variations of $\frac{b_y}{\theta}$ and $\frac{b_z}{\theta}$ with distance $\frac{x}{\theta}$ are well correlated as shown in Fig 11.23. The variation of the maximum excess velocity, $\frac{U_a}{\Delta u_{m0}}$, with distance $\frac{x}{\theta}$ is shown in Fig 11.24.

A mean curve can describe the experimental data for $\beta \leq \frac{1}{4.5}$ and it can be observed that u_{mo} values are lower than the coflowing free jet, especially for $\frac{x}{\theta} > 5$. The boundary shear stress, τ_0 at a given location x along the axis of the jet can be predicted from the plot of $\left(\frac{U_a}{u_*}\right)^2$ versus $\frac{x}{\theta}$, shown in Fig 11.25.

11.8 Summary

For plane wall jets $u_m \propto \frac{1}{\sqrt{x}}$, $b \propto x$, and the boundary shear stress $\tau_0 \propto \frac{1}{x}$. The distribution of the normalised velocity along a direction normal to the wall can be described by Verhoff's empirical equation. The coefficient of entrainment $\alpha_e = 0.035$, as given in Eq 11.10.

The growth of the three-dimensional bluff wall jets, including the circular wall jet, is quite different from that of free jets. The presence of the wall leads to anisotropy in the fluctuating components of the velocity. As a result, the jet spreads 4 to 5 times faster in the transverse direction, compared to the growth

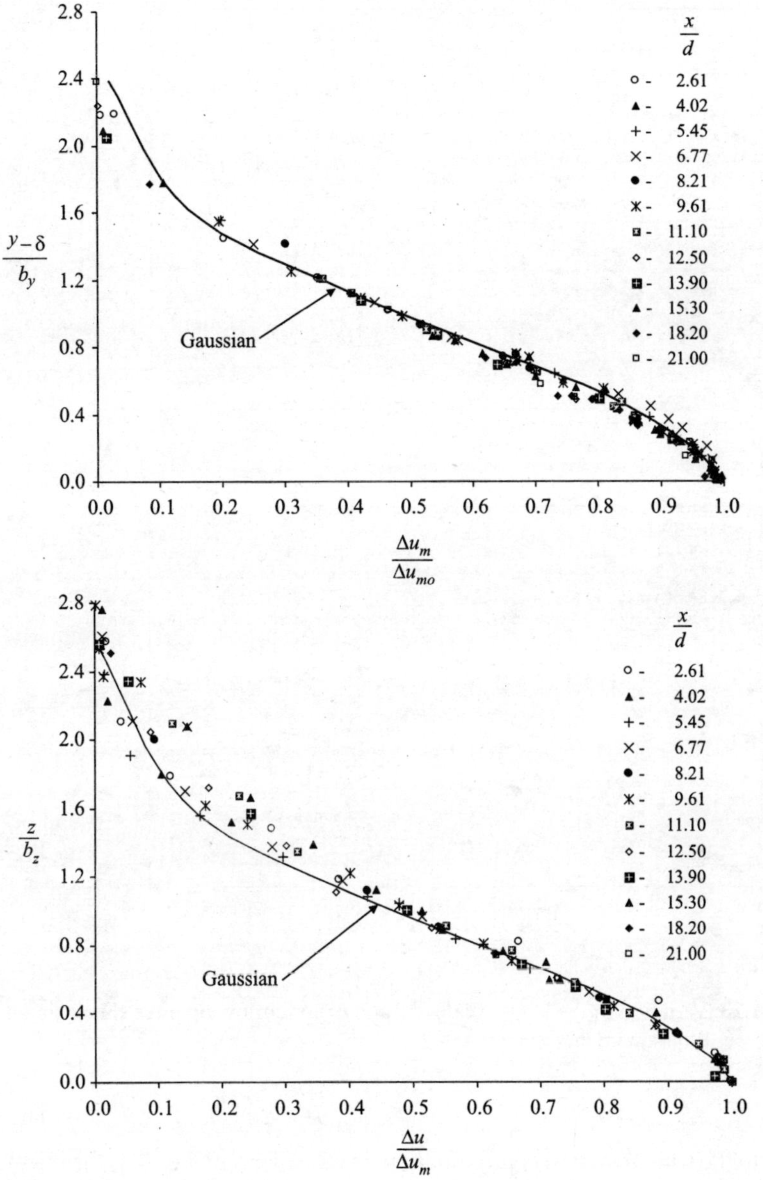

Figure 11.22 Distribution of the velocity-excess for coflowing wall jets (Source: Rajaratnam and Stalker, 1982)

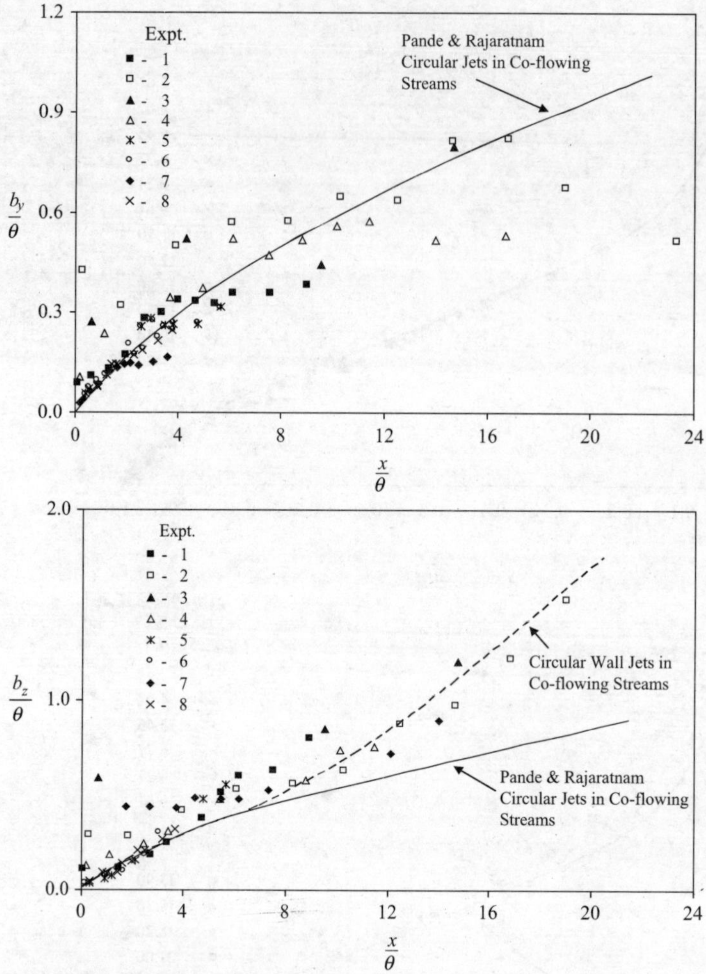

Figure 11.23 Variation of the length scales b_y and b_z for coflowing three-dimensional jets (Source: Rajaratnam and Stalker, 1982)

normal to the wall. As a first approximation, in case of circular outlets the effect of the boundary shear can be ignored for $\frac{x}{d} \leq 100$. Hence, one can consider $u_m \propto \frac{1}{x}$, and $b \propto x$. The local coefficient of friction c'_f on the axis attains a near constant value of 0.0065 for $\frac{x}{\sqrt{A}}$ exceeding a value of 40. The transverse distribution of the boundary shear stress τ_0, too, follows a Gaussian profile and exhibits self-similarity.

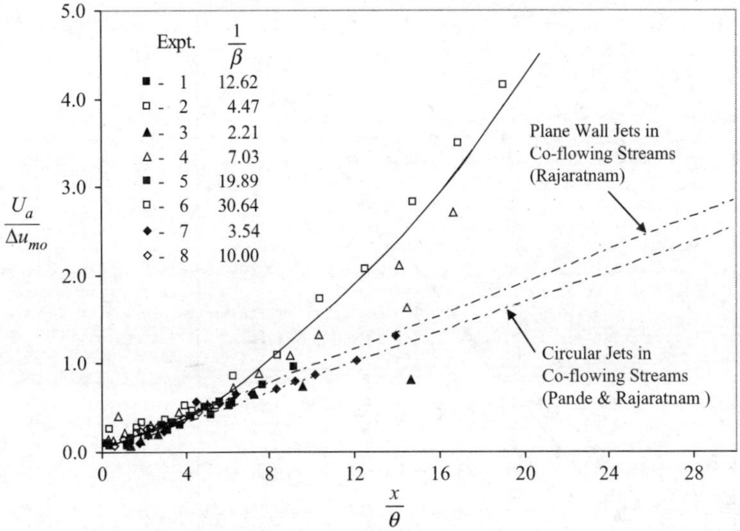

Figure 11.24 Decay of the maximum excess-velocity in a coflowing stream: Three-dimensional jets (Source: Rajaratnam and Stalker, 1982)

Based on Reichardt's hypothesis, the momentum flux from multiple outlets can be superposed to find the resultant distribution of the square of the velocities. The decay of the maximum momentum flux along the axis of symmetry of an array of jets can be represented in an unified manner if $\sqrt{NA}$ is used to normalize the distance x. For a single jet, the effect of the shape of the outlet can be taken care of by selecting $\sqrt{A}$ for normalizing the distance x.

The maximum concentration of a tracer in a wall jet decays faster than the maximum mean velocity. The length scales for concentration distribution, too, vary linearly with the distance x. The intensity of turbulence for three-dimensional wall jets lies somewhere between the values of the plane wall jet and the circular free jet.

In case of coflowing wall jets, the velocity-excess exhibits similarity. By using the momentum thickness θ as a normalising parameter, the variation of the length scales b_y and b_z with the distance x can be uniquely expressed independent of the velocity ratios.

Example 11.1: A plane wall jet, 10 cm wide, discharges air with an efflux velocity of 80 m/s. A two-dimensional sill, 20 cm in height, is held normal to

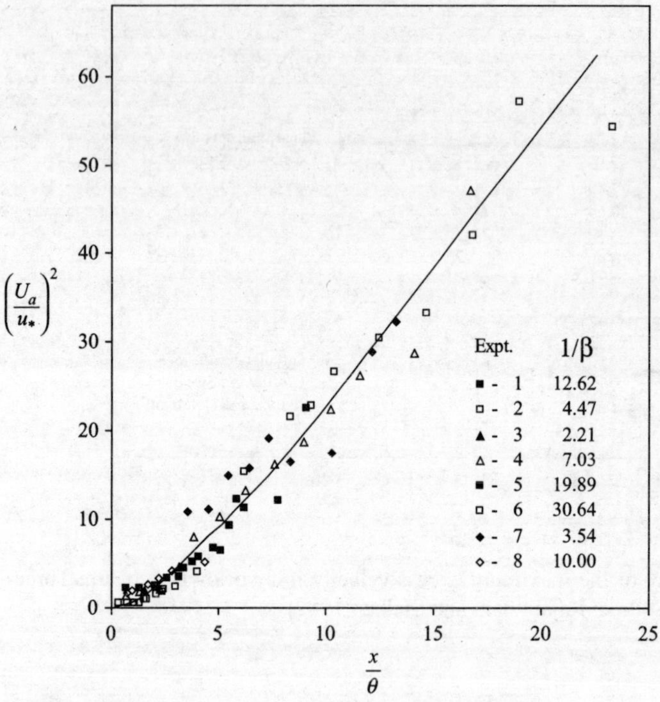

Figure 11.25 Variation of the boundary shear stress along the axis of a coflowing jet (Source: Rajaratnam and Stalker, 1982)

the air stream at a distance of 2.5 m from the nozzle. Find the force/unit length experienced by the sill.

You may adopt a simple model for the dynamic pressure acting on the front face based on Verhoff's wall jet velocity profile. Assume the wake pressure as $\left(-0.2\frac{\rho u_m^2}{2}\right)$ acting uniformly over the height of the sill.

Solution: Given, $b_0 = 0.10$ m, $u_0 = 80$ m/s, sill height $s = 0.2$ m, $\rho = 1.2$ kg/cubic-meter.

For the given location, $x = 2.5$ m, $\frac{x}{b_0} = \frac{2.5}{0.1} = 25$

$$\frac{u_m}{u_0} = \frac{3.5}{\sqrt{\frac{x}{b_0}}}, \text{ or } u_m = \frac{80 \times 3.5}{\sqrt{25}} = 56.0 \text{ m/s}.$$

The half-width, $b_{0.5} = 0.068 \times 2.5 = 0.17$ m: the thickness of the inner layer, $\delta = 0.16 \times b_{0.5} = 0.027$ m

Divide the sill height into 5 equal strips, 0.04 m thick. The jet velocity at the mid-point of each strip is evaluated from Verhoff's formula, Eq 11.1 and tabulated as follows.

y (m)	η	$\frac{u}{u_m}$	u (m/s)	$\rho \frac{u^2}{2} dA$(N)
0.02	0.1176	1.0	56.0	75.2
0.06	0.3529	0.91	51.0	62.4
0.10	0.5882	0.785	44.0	46.5
0.14	0.8235	0.617	37.5	33.7
0.18	1.0588	0.461	25.8	16.0
			total = 233.8(+)	

Wake suction force $= 0.2\left[\frac{\rho \times 56^2}{2}\right] \times 0.20 = 75.26$ N/m.

Hence, total force/unit length of the sill $= 233.8 + 75.26 \simeq 309$ N/m.

Example 11.2: The non-dimensional velocity profiles in the y and z directions are found to be self similar and are expressed as

$$\frac{u_m}{u_{m0}} = f_1\left(\frac{y}{b_y}\right) = f_1(\eta_y): \text{ normal to the boundary}$$

$$\frac{u}{u_m} = f_1\left(\frac{z}{b_z}\right) = f_2(\eta_z): \text{ parallel to the boundary}$$

Generate the isovels $\frac{u}{u_{m0}} = 0.5$, and 0.8. Plot the curves in the $\eta_y - \eta_z$ plane.

Solution: Making use of Eqs 11.11 and 11.12, for the functions f_1 and f_2, an expression for the velocity ratio $\frac{u}{u_{m0}}$ can be written as given by Eq 11.23. A spread sheet could be used to generate the values of the velocity ratio for different combinations of η_y and η_z. The required isovels are generated by interpolation. Typical plots are shown in the accompanying figure.

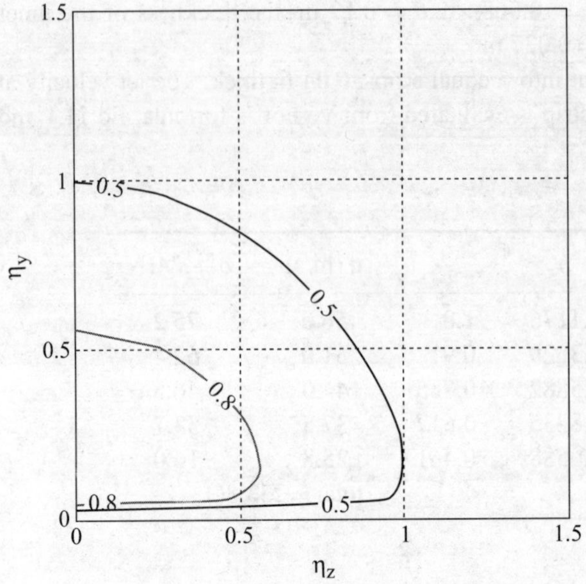

Isovels of a circular wall jet

Example 11.3: Effluent is being discharged into a pool of stagnant water as a wall jet. The outfall is a deeply submerged, rectangular sewer of size 1 m × 1.5 m, and the efflux velocity is 2 m/s . Find the minimum dilution and the spread of the jet at a distance of 20 m downstream of the outlet.

As an alternate option, it is proposed to discharge the effluent from 3 rectangular outlets, each 1 m high and 0.5 m wide, spaced at 2 m centre-to-centre. Estimate the minimum dilution at the above location. Contrast the result with the dilution that can be achieved had the jets been discharging as free jets.

Solution: Maximum concentration of the effluent will result in a minimum dilution. The distribution of concentration along y is given in Eq 11.27 as

$$\frac{C_m}{C_{mo}} = \exp\left(-\frac{y}{1.21 b_{yc}}\right)^2$$

Similarly, from Eq 11.28, the distribution of concentration in the z-direction is

$$\frac{C}{C_m} = \frac{1}{\left(0.965\frac{z}{y_{zc}}\right)^2 + 1}$$

For ease of computation, replace the rectangular nozzle by an equivalent circular outlet, $d = 1.38$ m.

From Eq 11.30, $b_{yc} = 0.061x + 0.31d = 0.061 \times 20 + 0.31 \times 1.38 = 1.65$ m.

Similarly, from Eq 11.31, $b_{zc} = 0.31x - 3d = 0.31 \times 20 - 3 \times 1.38 = 2.06$ m.

The variation of the maximum concentration with distance-x can be found from Eq 11.29. Thus,

$$\frac{C_{mo}}{C_0} = \frac{6.97}{\frac{20.0}{1.38}} = 0.48$$

The minimum dilution, $\frac{C_0}{C_{mo}} = \frac{1}{0.48} \simeq 2.$

Average dilution $\overline{S} = 1.7 \times 2 = 3.4.$

The edge of the jet lies at a distance of 2.2 times the half width. Hence, in plan, the edges will be separated by a distance $= 2 \times (2.2 \times 2.06) = 9.0$ m.

In contrast, normal to the wall, the jet will spread up to a distance of $2.2 \times 1.65 = 3.64$ m.

Let us consider the case of three jets. The area of each outlet, $A = 0.5 \times 1.0 = 0.5\ \text{m}^2$.

The equivalent diameter $d = \sqrt{\frac{4 \times 0.5}{\pi}} = 0.798$ m.

Corresponding to $x = 20m$, $\frac{x}{d} = \frac{20.0}{0.798} = 25.06.$

Adopting Law and Herlina's expression for the maximum velocity u_{mo},

$$\frac{u_{mo}}{U_0} = 9.23\left(\frac{x}{d}\right)^{-1.066} = \frac{9.23}{25.06^{1.066}} = 0.298.$$

From Eq 11.29,

$$\frac{C_{mo}}{C_0} = \frac{6.97}{\left(\frac{x}{d}\right)} = \frac{6.97}{25.06} = 0.278.$$

Hence, for an individual jet, $\frac{C_{mo}u_{mo}}{C_0 U_0} = 0.278 \times 0.298 = 0.0828$

The width of the concentration profile in the z-direction, $b_{zc} = 0.31x - 3d = 0.31 \times 20.0 - 3 \times 0.798 = 3.806$ m.

Similarly, from Eq 11.21, the width of the velocity profile, $b_z = 3.0$ m.

At a distance of $z = 2.0$ m from any one of the jet axis,

$$C_m u_m = C_{mo} u_{mo} \left\{ \frac{1}{\left(0.965 \frac{z}{b_{zc}}\right)^2 + 1} \right\} \left\{ \frac{1}{\left(0.965 \frac{z}{b_z}\right)^2 + 1} \right\}$$

Substituting for z, b_{zc} and b_z, $C_m u_m = 0.562 C_{mo} u_{mo}$. By superposing the $(C_m u_m)$ values of the three jets, on the axis of symmetry,

$$\sum (C_m u_m) = (C_m u_m)_3 = C_{mo} u_{mo} + 2 \times 0.562 C_{mo} u_{mo} = 2.125 C_{mo} u_{mo}.$$

Again, making use of the Gaussian distribution, the square of the velocity, u_m^2 at a distance of $z = 2.0$ m, can be found as $u_m^2 = 0.627 \times u_{mo}^2$.

By the method of superposition,

$$\sum u_m^2 = \left(u_m^2\right)_3 = u_{mo}^2 + 2 \times 0.627 u_{mo}^2 = 2.253 u_{mo}^2.$$

From the above, $(u_m)_3 = 1.5 u_{mo}$.
Hence,

$$(C_m)_3 = \frac{2.125 \times (C_{mo} u_{mo})}{(u_m)_3} = 1.416(C_{mo}) = 1.416 \times 0.278 C_0 = 0.394 C_0$$

Thus, the minimum dilution at a distance of $x = 20$ m is, $\frac{1}{0.394} = 2.54$. As expected, this value is somewhat higher than the value for the single large jet of size 1.5 m × 1.0 m.

Example 11.4: A wall jet is used as a means of measuring the attachment strength of zebra mussels. A fluid detachment parameter (DP) is defined as the nominal wall shear stress at detachment site, multiplied by the square of the mussel length. The DP varies significantly with the nature of the substrate. Information of this kind could be useful for environmentally favorable zebra mussel control.

A 2 cm × 3 cm rectangular duct is used to generate a wall jet on a PVC substrate. An isolated mussel, 10 mm in diameter has a specific gravity of 1.1, and calculations show that its passive resistance motion is negligible. For a jet velocity of 5.5 m/s, the mussel detaches from the substrate. Find the DP if the mussel is located at a distance of 30 cm downstream of the nozzle and is 3 cm away from the centreline of the jet.

Solution: Given, $x = 30$ cm, $z = 3.0$ cm, and $A = 3.0 \times 2.0 = 6.0 cm^2$
Hence, $\sqrt{A} = \sqrt{6.0} = 2.45$ cm: $\frac{x}{\sqrt{A}} = \frac{30.0}{2.45} = 12.25$
From Eq 11.24, with $C = 0.122$,

$$\frac{u_{mo}}{U_0} = \frac{0.94}{0.122}\left(\frac{x}{\sqrt{A}}\right)^{-1} = 0.629$$

Thus, the maximum velocity u_{mo}, at $x = 30.0$ cm and $z = 0$, is $(0.629 \times 5.5) =$ 3.14 m/s.

From Fig 11.6, $c_f' = 1.8 \times 10^{-3}$ corresponding to $\frac{x}{\sqrt{A}} = 12.25$.

Hence, the centreline boundary shear $\tau_{mo} = c'_f \times \left(\frac{1}{2}\rho U_0^2\right) = \left(1.8 \times 10^{-3}\right) \left(\frac{1}{2} \times 1000 \times 5.5^2\right) = 27.2\ \text{N/m}^2$.

The boundary shear stress τ_o at a distance of $z = 3$ cm away from the jet axis can be found from Fig 11.8. For $\frac{z}{x} = 0.1$, the ratio $\frac{\tau_o}{\tau_{mo}} = 0.5$. Hence, the boundary shear stress $\tau_o = 0.5 \times 27.2 = 13.6\ \text{N/m}^2$.
Diameter of the mussel $= 10$ mm $= 10^{-2}$ m.
The surface area of the mussel, $A = \frac{\pi}{4} \times 10^{-4} = 0.785 \times 10^{-4}\ \text{m}^2$.
Therefore, the detachment parameter, $DP = 13.6 \times 0.785 \times 10^{-4} = 1.067 \times 10^{-3}$ N.

Example 11.5: Surplus water from a small reservoir is being discharged from a circular outlet 0.3 m in diameter. The efflux velocity is 20 m/s and the discharge takes place tangential to the bottom of an after-bay. The tailwater is significantly large and the outlet can be treated as deeply submerged. Find the following,

(a) The distribution of the mean forward velocity in the x–y plane of symmetry at a distance of 6.0 m from the outlet,
(b) The transverse distribution of the boundary shear stress τ_0, and
(c) The thickness and width of the forward moving shear layer (the jet).

Treat the downstream channel as wide and smooth. What will be the consequence if the conveyance channel is narrow?

Solution: Diameter of the outlet, $d = 0.3$ m, and $U_0 = 20$ m/s.
$A = \frac{\pi}{4} \times 0.3^2 = 0.0707\ \text{m}^2$.
From Eq 11.10,

$$\frac{u_{mo}}{U_0} = \frac{0.94}{C \times \left(\frac{x}{\sqrt{A}}\right)} = \frac{0.94}{0.122 \times 22.5} = 0.342$$

Hence, the maximum velocity $u_{mo} = 0.342 \times 20.0 = 6.84$ m/s.
The half width of the jet in the y-direction, $\frac{b_y}{d} = 0.90 + 0.045(\frac{x}{d}) = 0.90 + 0.045 \times 20.0 = 1.85$. Thus, $b_y = 1.85 \times 0.3 = 0.555$ m.
The thickness of the inner layer $\delta = 0.16 \times b_y = 0.089$ m.
From Verhoff's relationship, Eq 11.1,

$$\frac{u_m}{u_{mo}} = 1.48\eta^{\frac{1}{7}}\{1 - erf\ (0.68\eta)\}$$

The magnitudes of the forward velocity u_m found from the above expression are as follows:

y(m)	η	u_m (m/s)
0.0	0.0	0.0
0.02	0.036	6.11
0.089	0.16	6.84
0.15	0.27	6.68
0.30	0.54	5.61
0.50	0.90	3.85
0.75	1.35	2.07
1.00	1.80	0.91

From Fig 11.7, the coefficient of skin-friction, $c_f = \frac{\tau_{om}}{\frac{1}{2}\rho u_{mo}^2} = 4.5 \times 10^{-3}$

Hence, $\tau_{om} = (4.5 \times 10^{-3})\left(\frac{1}{2} \times 1000 \times 6.84^2\right) = 105.26 N/m^2$.
$\frac{b_\tau}{d} = 0.093\frac{x}{d}$. Hence, $b_\tau = 0.093 \times 6.0 = 0.558$ m.
Thus, the distribution of the boundary shear stress along z-direction is given as

$$\frac{\tau_o}{\tau_{mo}} = \mathrm{e}^{-0.693\left(\frac{z}{b_\tau}\right)^2}$$

From Eq 11.20, the half width in the y direction is 0.555 m. Similarly, from Eq 11.21, the half-width in the z direction is given as

$$\frac{b_z}{d} = 0.20\left(\frac{x}{d}\right) - 1.25$$

For the given outlet, at a location of $x = 6.0$ m, the half-width in the z-direction is 0.825 m. Let us assume that the edge of the jet is located at a distance of 2.2 times the b_y and b_z values. Hence, the thickness of the jet normal to the wall $= 2.2 \times b_y = 2.2 \times 0.555 = 1.22$ m.
The total jet width in the z-direction $= 2 \times (0.825 \times 2.2) = 3.63$ m.

In case of a narrow conveyance channel, the side walls will act as reflectors and the momentum flux from the imaginary sources needs to be superposed. Consequently, larger values for the maximum velocity will result. If the side walls are too close, the flow can behave like a plane wall jet and $u_{mo} \propto x^{-\frac{1}{2}}$.

Example 11.6: Three deeply submerged sluices, 0.6 m(*wide*) × 1.0 m(*high*), are spaced at 1.0 m centre-to-centre. They discharge water downstream of the reservoir with a velocity of 20 m/s near the channel bottom. A stilling basin is provided to dissipate the kinetic energy of the jets. The beginning 10 m length of the basin is of concrete, followed by stone rip-rap protection.

Find the maximum shear stress acting on the stones and recommend the minimum size of stone to be used in the design. Make suitable assumptions.

Solution: Given, $U_0 = 20$ m/s, $A = 0.6\text{ m} \times 1.0\text{ m} = 0.6\text{ m}^2$, and $\sqrt{A} = 0.774$ m.

Let us assume that the outlets are deeply submerged and the concrete floor is a smooth boundary.

$x = 10.0$ m, and the corresponding $\frac{x}{\sqrt{A}} = \frac{10.0}{0.774} = 12.9$.

From Eq 11.24, $\frac{u_{mo}}{U_0} = \frac{0.94}{C}\left(\frac{x}{\sqrt{A}}\right)^{-1}$, where $C = 0.122$.

Thus, $u_{mo} = 20 \times \frac{0.94}{0.122} \times \frac{1}{12.9} = 11.94$ m/s

The velocity profiles in the y and z directions are assumed to be self similar. Hence,

$$\frac{u_m}{u_{mo}} = \mathrm{e}^{-0.693\eta_z^2}; \quad \frac{u_m^2}{u_{mo}^2} = \mathrm{e}^{-1.386\eta_z^2}$$

To find the maximum velocity along the axis of the middle outlet, let us make use of the method of superposition of the square of the velocities. The equivalent diameter of each outlet, $d = 0.874$ m. At the desired location $x =$ 10.0 m, the half width of the jet, b_z is evaluated as follows:

$$\frac{b_z}{d} = 0.20\left(\frac{x}{d}\right) - 1.25 = 0.20 \times \frac{10.0}{0.874} - 1.25 = 1.038$$

Hence, $b_z = 1.038 \times d = 1.038 \times 0.874 = 0.907$ m.

The centre to centre spacing is 1.0 m. Thus, by letting $z = 1.0$ m, the non-dimensional distance, $\eta_z = \frac{z}{b_z} = \frac{1.0}{0.907} = 1.10$. Considering the u^2 contributions of the three jets, we have

$$\left(u_{m0}^2\right)_3 = \left(u_{m0}^2\right)\left\{1 + 2 \times \mathrm{e}^{-1.386\eta_z^2}\right\} = 142.56\left\{1 + 2 \times \mathrm{e}^{-1.386\times 1.10^2}\right\}$$
$$= 195.54\ \mathrm{m}^2/s^2.$$

Hence, the maximum velocity at $x = 10.0$ m on the axis of symmetry $= \sqrt{195.54} = 13.98$ m/s.

From Fig 11.7, the coefficient of skin friction $c_f' = 4.5 \times 10^{-3}$ corresponding to $\frac{x}{\sqrt{A}} = 12.9$.

Therefore, the maximum boundary shear $\tau_{0m} = \left(4.5 \times 10^{-3}\right) \times \left(\frac{1}{2} \times 1000 \times 13.98^2\right) = 439.74\ \mathrm{N/m}^2$

From Shields diagram on initiation of motion of sediment particles, for shear Reynolds number exceeding 400, the critical diameter (i.e. the minimum size) of the stone can be found from the relationship

$$d_c = \frac{\tau_0}{0.056(\gamma_s - \gamma)} = \frac{439.74}{0.056 \times (1000 \times 9.81)(2.6 - 1.0)} = 0.50 \text{ m}$$

Example 11.7: Taking the cue from coflowing free jets, it is proposed to discharge effluent from a circular outlet as a coflowing wall jet. The mean stream velocity is 1 m/s and the efflux velocity of the jet at the outlet is 5 m/s. If the diameter of the circular outlet is 0.3 m, find the minimum dilution at downstream distances of 5, 10 and 20 m from the outlet. Contrast these results with a free jet having identical flow conditions and comment.

Solution: Given, $U_0 = 5.0$ m/s, $U_a = 1.0$ m/s. Hence, $\delta U_0 = (U_0 - U_a) =$ 4.0 m/s

For $x = 5.0$ m, the non-dimensional distance $\frac{x}{d} = \frac{5.0}{0.3} = 16.33 < 20$. Yet, we shall assume that the following expression for the maximum concentration is valid for $x = 5.0$ m location as well.

$$\frac{C_{mo}}{C_0} = \frac{6.97}{\left(\frac{x}{d}\right)}$$

The ratio of the maximum concentration and the corresponding (minimum) dilution at different x-locations are shown in the following table.

x,(m)	$\frac{x}{d}$	$\frac{C_{mo}}{C_0}$	Min. Dilution
5.0	16.33	0.427	2.34
10.0	33.33	0.209	4.78
15.0	50.0	0.139	7.17

For a coflowing free circular jet, the minimum dilution occurring on the axis of the jet is

$$S_c = \frac{C_0}{C_m} = \frac{\lambda^2 \pi B^2}{2Q_0}\left(U_a + \frac{2}{1+\lambda^2}\delta U\right)$$

The excess velocity δU and the jet width B are functions of the distance $\frac{x}{\theta}$, refer Figs. 6.2 and 6.7. Assuming $\lambda = 1.2$, one can find the minimum dilution S_c from the above expression. However, one can also use Fig 6.8 to find S_c.

In either case we need to know the momentum thickness θ. In terms of the velocity ratio $\beta = \frac{U_a}{U_0}$ we can find the momentum thickness as,

$$\theta = \frac{d}{2}\sqrt{\frac{1}{\beta}\left(\frac{1}{\beta}-1\right)} = \frac{0.3}{2}\sqrt{5\times 4} = 0.67 \text{ m}$$

As a typical example, for $x = 10$ m, $\frac{x}{\theta} = 14.9$.

From Fig 6.8, the corresponding value of $S_c\left(\frac{\beta}{1-\beta}\right)$ is approximately equal to 2.25. Hence the dilution, $S_c = \frac{2.25}{\frac{0.2}{(1-0.2)}} = 8.9$. Thus, the attainable dilution in case of a free jet is nearly twice as much as that of the wall jet.

12 Comparison with CFD Analysis

12.1 Introduction

CFD Softwares are often used in the analysis and design of fluid systems or devices. They simulate flows of gases and liquids, heat and mass transfer, etc., representing the required system or device. CFD is a powerful predictive tool for assessing the manner in which the flow will behave for a set of imposed boundary conditions. Changes can be made quickly in the design to arrive at an optimal solution. Physical models are, nonetheless, tested to check the actual performance of the design. The range of physical models for testing is narrowed down via CFD simulations and considerable savings in time and other resources can result.

Significant progress has taken place in all the three aspects of the CFD software, namely, preprocessing, solving and post-processing. Of the many commercially available softwares, designers and modelers of the fluid system use FLUENT and Flow3D codes. Often the modeler has some flexibility in generation of the grid, its size, imposition of the boundary conditions, selection of the size of the domain and choice of coefficients in a turbulence model. Therefore, for arriving at meaningful solutions, trial runs are essential. Making use of experimental data, the numerical models are calibrated, especially for well established cases of flow.

CFD results of simple jets are included in this chapter and the predicted mean flow features are compared with the results of the point-source method discussed earlier in this book. The standard $k - \varepsilon$ turbulence model is usually selected in the CFD simulation without attempting to fine tune the answers by repeating the trial runs.

12.2 Plane Free Jet

The plane turbulent free jet is the simplest free shear flow which has been studied experimentally as well as theoretically. The flow field of the same has been simulated by using different types of turbulence models (Anderson and Spall, 2001). Based on the point-source method, the decay of the maximum specific momentum flux for a three dimensional outlet was presented in Chapter 5 (Eq 5.53) as

$$\frac{u_{m0}}{U_0} = \left\{ erf\left(\frac{B_0}{\sqrt{2}b}\right) erf\left(\frac{L_0}{\sqrt{2}b}\right) \right\}^{\frac{1}{2}} \tag{12.1}$$

For a plane jet, $L_0 \to \infty$ and the above expression simplifies to

$$\frac{u_{m0}}{U_0} = \left\{ erf\left(\frac{B_0}{\sqrt{2}b}\right) \right\}^{\frac{1}{2}} \tag{12.2}$$

It is customary to use the width of the slot b_0 in the velocity decay formula of a plane free jet and denote the maximum velocity on the axis as u_m. Since, $B_0 = \frac{b_0}{2}$ and the width scale $b = c \times x$, Eq 12.2 can be further simplified by assigning a value of $c = 0.0565$ and the final theoretical expression is

$$\frac{u_m}{U_0} = \left\{ erf\left(\frac{6.257}{\frac{x}{b_0}}\right) \right\}^{\frac{1}{2}} \tag{12.3}$$

After analyzing experimental data from different sources, the following empirical expression has been recommended for $\frac{x}{b_0} \geq 6.1$ (Rajaratnam, 1976)

$$\frac{u_m}{U_0} = \frac{2.475}{\sqrt{\frac{x}{b_0}}} \tag{12.4}$$

A finite-volume code FLUENT has been used to solve the plane jet problem based on the two-dimensional, Reynolds-averaged Navier-Stokes equations (Anderson and Spall, 2001). Both Reynolds stress (RSM) and the $\kappa - \varepsilon$ models were used for turbulence closure. The computational domain was a rectangular region, bounded on the $x = 0$ plane by a solid wall along which standard equilibrium wall functions were specified. A symmetry boundary condition was defined along the $y = 0$ plane. A uniform inlet velocity was imposed at the efflux section and the turbulence intensity at the inlet was set to 3.6 per cent. The CFD results for both the turbulence models were nearly the same and agreed with the experimental observations recorded by a hot wire anemometer as shown in Fig 12.1.

The point-source results from Eq 12.3 are superposed in the figure and are seen to be in good agreement with the CFD results over the entire range of distances shown. Nonetheless, differences persist between the two predicted curves and the mean experimental curve.

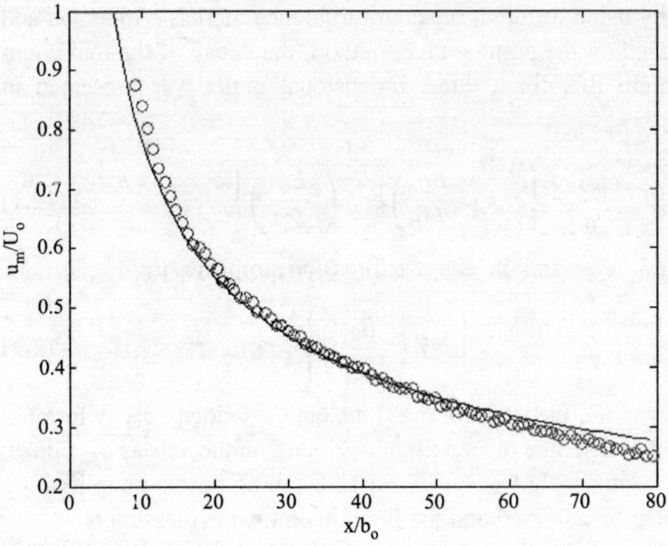

Figure 12.1 Decay of the maximum velocity in a plane free jet (Source: Anderson and Spall)

With the point-source method, two important advantages accrue, namely the spread coefficient c required to find b in Eq 12.2 is identical to the axisymmetric case and the predicted u_m values are very nearly the same as the CFD results. The simulation results exhibit other features of the plane turbulent jet, namely, self-similarity of the velocity profiles in the developed flow region and a linear growth of $b_{0.5}$ and b_g with distance x. The major limitation of the CFD results is that the rate of spread of the jet with x is significantly larger than the experimental values.

12.3 Plane Wall Jet

For a plane turbulent wall jet on a smooth boundary under zero pressure gradient, the velocity distribution normal to the boundary (i.e. the wall) is described by the classical wall jet profile as in Eq 11.1. In addition, based on the observed data of various researchers, an average decay curve for the maximum velocity was proposed as (Rajaratnam, 1976)

$$\frac{u_m}{U_0} = \frac{3.50}{\sqrt{\frac{x}{b_0}}} \qquad (12.5)$$

where b_0 is the width of the slot and U_0 is a uniform velocity at the efflux section.

Herein, a simple superposition method is used to predict the decay of the maximum velocity of a wall jet based on the results of the free jet. For example, in case of a plane turbulent free jet, the following has been proposed (Lee and Chu, 2002),

$$\frac{u_m}{U_0} = \frac{2.58}{\sqrt{\frac{x}{b_0}}} \tag{12.6}$$

The wall can be considered as a frictionless boundary so that the mirror image of the slot of width b_0 is included in predicting the maximum velocity u_m of a wall jet. Therefore for a plane wall jet of width b_0, Eq 12.6 can be modified by substituting $2b_0$ in lieu of b_0 resulting in

$$\frac{u_m}{U_0} = \frac{3.648}{\sqrt{\frac{x}{b_0}}} \tag{12.7}$$

Following a similar procedure, Eq 12.3 (point-source method) can be modified for a plane wall jet to express the variation of u_m as

$$\frac{u_m}{U_0} = \left\{erf\left(\frac{12.514}{\frac{x}{b_0}}\right)\right\}^{\frac{1}{2}} \tag{12.8}$$

Interestingly, a similar transformation in the widths of the nozzles by a factor of 2 also links Eqs 12.4 and 12.5.

The expression of Eq 12.7 is valid beyond the potential core region. In contrast, Eq 12.8 holds good for the entire range of distances starting from the efflux section. Typical values of $\frac{u_m}{U_0}$ from the two equations, Eqs 12.7 and 12.8, are shown in the following Table.

From Table 12.1, it can be noticed that in the fully developed region of the wall jet, the difference between the values from the two equations is very small and is of the order of 2 per cent and could be due to the fact that the wall friction is ignored in Eq 12.8. Significant differences are noticed between the two in the developing zone, $\frac{x}{b_0} \leq 15$. Nonetheless, the predictions from the point-source lie closer to the experimentally observed data.

To contrast the decay of the maximum velocity predicted by the point-source technique with CFD simulation results, computations were carried out for an air jet of width $b_0 = 0.005$ m, having an efflux velocity $U_0 = 15.0$ m/s. The Reynolds number exceeds 4500 and the flow can be treated as turbulent. Results pertaining to the variation of $\frac{u_m}{U_0}$ with distance $\frac{x}{b_0}$ is shown in Fig 12.2.

Table 12.1 Plane wall jet velocity decay

$\frac{x}{b_0}$	$\frac{u_m}{U_0}: EQ.(12.7)$	$\frac{u_m}{U_0}: EQ.(12.8)$
4	1.0	0.999
10	1.0	0.956
15	0.942	0.873
25	0.730	0.721
40	0.577	0.584
60	0.471	0.481
100	0.365	0.375
150	0.298	0.308
200	0.258	0.265

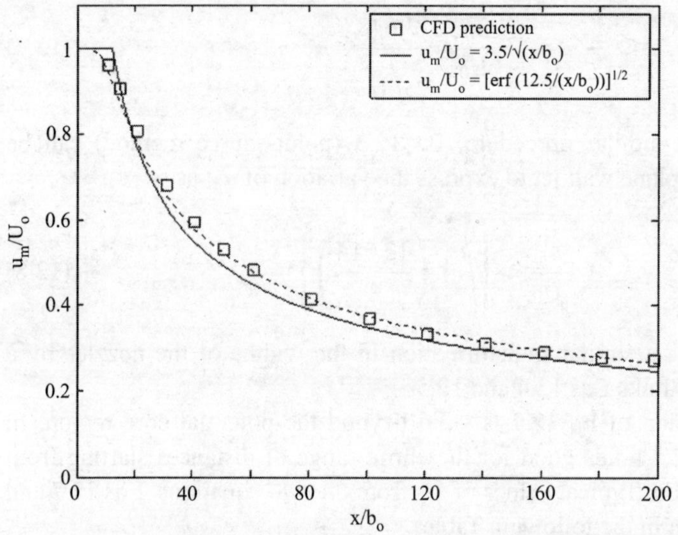

Figure 12.2 Decay of the maximum velocity in a plane wall jet

The CFD results are seen to be in good agreement with the values of the point-source method.

Other features of the planc wall jet from simulation include the velocity distribution at different cross-sections and the rate of growth of the half-velocity length scale $b_{0.5}$. Figure 12.3 shows that the velocity profiles are self-similar and are akin to the classical wall jet profile described by Verhoff's empirical relationship, Eq 11.1 shown as a dotted line. The maximum velocity u_m occurs at a distance of about $0.1 \times b_{0.5}$ for all the cross-sections of the

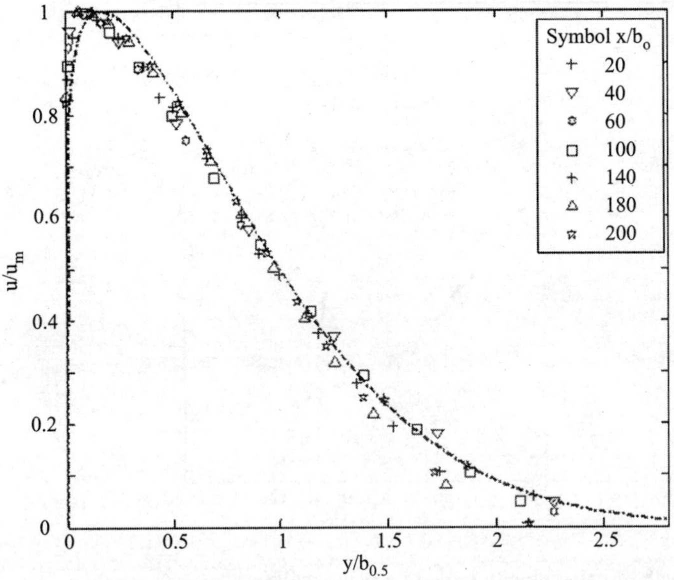

Figure 12.3 Self-similarity of velocity profiles of a plane wall jet

developed flow region. This is in variance from the commonly observed distance of $\delta = 0.16 \times b_{0.5}$ by several researchers. Furthermore, the simulation results in Fig 12.4 show that the half-width of the jet varies linearly with the distance and $b_{0.5} = 0.092x$. In contrast, experimental results of plane wall jets show that on an average $b_{0.5} = 0.068x$.

12.4 Axially Symmetric Free Jet

Analytical expressions have been worked out for this classical case based on Prandtl's theory of turbulence (Chapter 3). Reichardt's momentum transport theory was also adopted as an alternative and the velocity distribution was predicted as Gaussian (Chapter 4). Following the superposition of the momentum flux from point-sources spread over the outlet, an expression for the decay of the maximum velocity u_m was derived in Chapter 6 (Eq 6.46) as

$$\frac{u_m^2}{U_0^2} = 1 - \exp\left\{-\frac{1}{4}\left(\frac{d}{\sqrt{2}cx}\right)^2\right\} \tag{12.9}$$

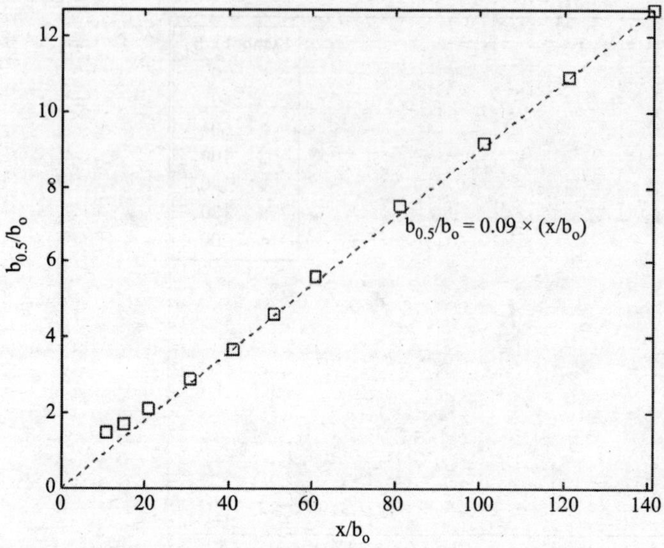

Figure 12.4 Growth of half-velocity scale of a plane wall jet

where the spread coefficient $c = 0.0565$. By substituting the value for c, the above expression can be rewritten as

$$\frac{u_m}{U_0} = \left(1 - \exp\left\{-\left(\frac{6.24}{\frac{x}{d}}\right)^2\right\}\right)^{\frac{1}{2}} \tag{12.10}$$

For large values of $\frac{x}{d}$, the small order terms in the series expansion of the exponential function of Eq 12.10 can be neglected, resulting in the relationship

$$\frac{u_m}{U_0} = \frac{6.24}{\frac{x}{d}} \tag{12.11}$$

Equation 12.11 is identical to the expression proposed by Lee and Chu (2003). It can be used only in the developed zone of a circular jet whereas Eq 12.10 is valid throughout the jet trajectory.

A commonality exists between CFD simulation and the point-source method since in the computations, both do not consider the developing and the developed flow regions of the jet as separate entities. Therefore, it will be interesting to contrast the results of the two approaches for a classical circular jet.

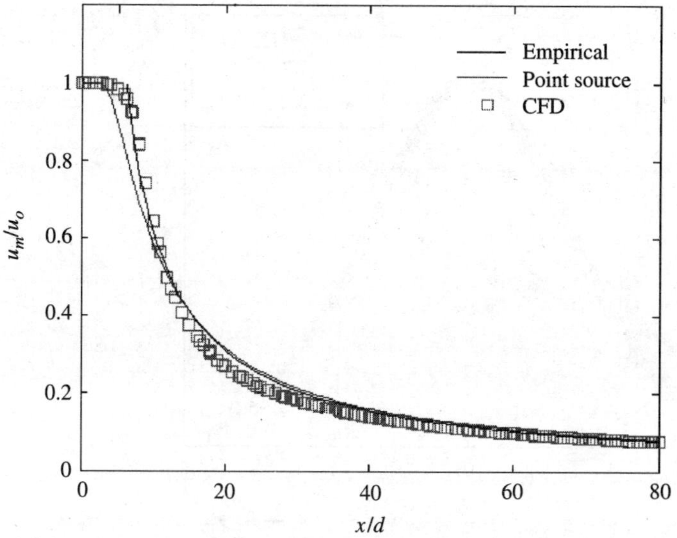

Figure 12.5 Decay of maximum velocity of an axially symmetric jet

A circular jet 10 mm in diameter and with an efflux velocity of 1.0 m/s was adopted to carry out the flow simulation. The results re-confirmed that the velocity profiles in the fully developed region of the jet satisfy self-similarity and the distribution is Gaussian. The characteristic width b_g was found to be a linear function of the distance x and can be expressed as $b_g = 0.124x$. The predicted spread rate is significantly different from the experimental value of 0.114 reported by researchers.

The predicted decay of the maximum velocity u_m with x from different methods is shown in Fig 12.5.

Equation 12.11 is included in the figure as it has been validated based on experiments. Beyond $\frac{x}{d} \simeq 10$ there is very little difference between the values from the point-source method and Eq 12.11. There are differences between the point-source and CFD results, especially for $15 \leq \frac{x}{d} \leq 35$. This could be a manifestation of the larger rate of spread in the simulation.

12.5 Three-dimensional Free Jets

The three-dimensional jet represents a more general case as the jet spreads at different rates along the y and z directions. Flow of water from a rectangular outlet of size 20 mm wide and 10 mm high was simulated for an efflux velocity

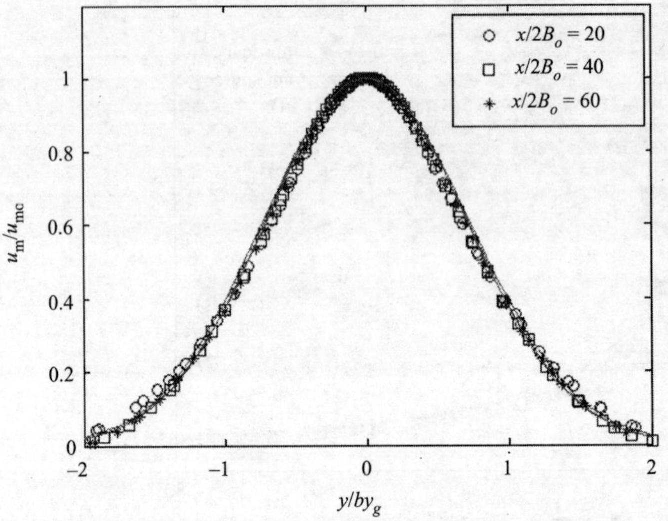

Figure 12.6 Mean velocity profiles of a three-dimensional jet

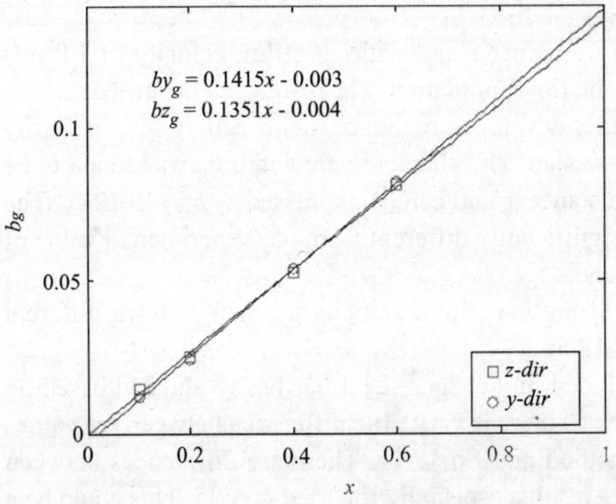

Figure 12.7 Growth of velocity scale of a three-dimensional bluff jet

of $U_0 = 1.0$ m/s. The simulated velocity profiles satisfy self-similarity and are Gaussian in the axially symmetric zone. A typical plot for the distribution of velocity u_m in the y direction is shown in Fig 12.6. The profiles in the z direction satisfy self-similarity beyond a larger x and for sufficiently large values of $\frac{x}{d}$ the distributions in the y and z directions look alike. The simulated growth of the jet in the y and z directions is shown in Fig 12.7, and the jet

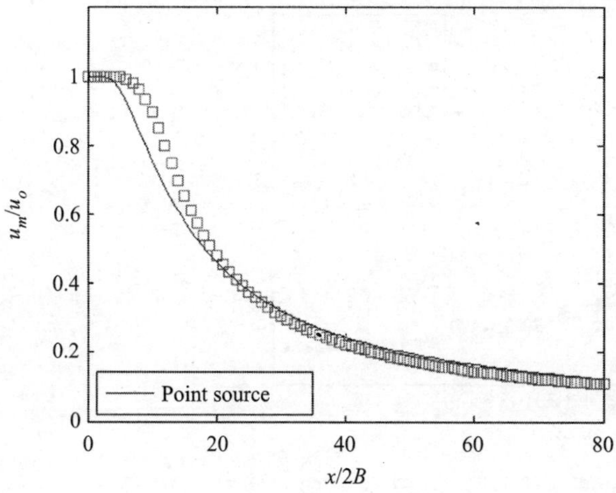

Figure 12.8 Decay of maximum velocity of a three-dimensional bluff jet

being a bluff jet, growth rates in the two mutually perpendicular directions are nearly the same. The predicted growth rate is larger compared to the observed values.

The decay of the maximum velocity $\frac{u_{mo}}{U_0}$ along the axis of the jet is depicted in Fig 12.8 indicating that the CFD and point-source results from Eq 12.1 are very much the same for $\frac{x}{2B_0} > 15$. In the developing flow region, significant differences exist between the values predicted by the two approaches.

12.6 Circular Jet in Shallow Ambient

In many practical problems, the circular jet discharges in shallow water. The downstream decay of the maximum velocity, mean velocity distribution across the jet and the spread of the jet can be assessed based on the CFD simulation accounting for the presence of the channel bottom and the free surface. The same mean flow features of the jet can be predicted based on the point-source method. For the purpose of illustration, a nozzle of diameter 12.7 mm and an efflux velocity of 3.437 m/s (used in the experiments of Raiford and Khan, 2009) are considered in the analysis. The computational domain is a rectangular tank of dimensions 1.27 m $(x) \times 0.1016$ m$(y) \times 1.2$ m(z) and the submergence depth $\frac{y_o}{d} = 4$. The full Reynolds-Averaged Navier-Stokes equations were solved using the realizable $k - \varepsilon$ model as turbulence closure.

The velocity profiles of the CFD simulation satisfied self-similarity, though a bit different from the Gaussian distribution. The CFD and the point-source

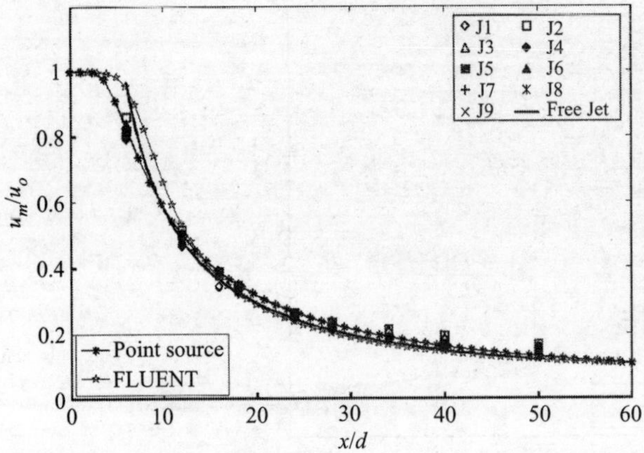

Figure 12.9 Decay of maximum velocity of a circular jet in shallow water

results for the decay of the maximum velocity u_m are given in Fig 12.9 and fair agreement between the two exists in the fully developed region of the jet. In contrast, once again in the developing flow region ($\frac{x}{d} \leq 10$) some differences can be observed between the two. The free jet relationship Eq 12.11 that ignored the effects of the free surface and the channel bottom, is also shown in the figure. The point-source method agrees with the data quite well over the entire range of distances covered in the experimental investigation. It is also clear that the effects of the free surface and the bottom are negligible for submergence ratio $\frac{y_o}{d} \geq 4$.

The effect of the submergence ratio has been investigated for a smaller value of $\frac{y_o}{d} = 2$ making use of the point-source method and the method of superposition of the momentum flux. Figure 12.10 compares the results from two different submergence ratios, and it can be seen that for $\frac{x}{d} \leq 25$ the free surface and the bottom have negligible effect. Beyond the said distance, the maximum velocity attains higher values for the $\frac{y_o}{d} = 2$ case and there is good agreement between the predicted and the experimental values (Raiford and Khan, 2009).

The velocity profiles in the surface zone have been predicted by superposing the momentum flux and typical results corresponding to $\frac{y_o}{d} = 4$ are depicted for two downstream locations in Fig 12.11. The good agreement between the predicted and the observed data reaffirm the efficacy of the point-source technique in capturing the mean velocity field of jets diffusing in an ambient fluid of limited depth.

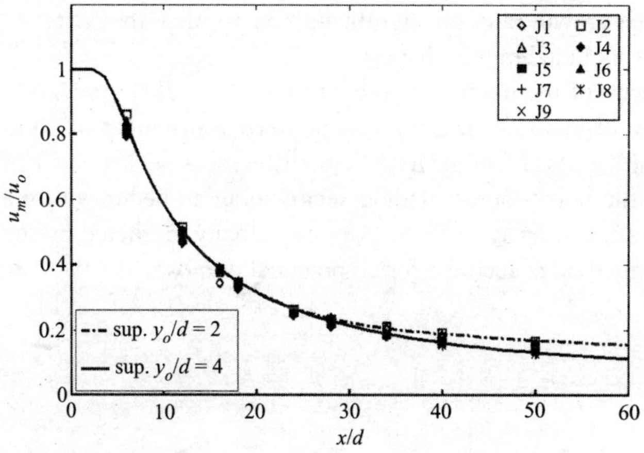

Figure 12.10 Effect of submergence on decay of maximum velocity (Source: Pani and Lai, 2010)

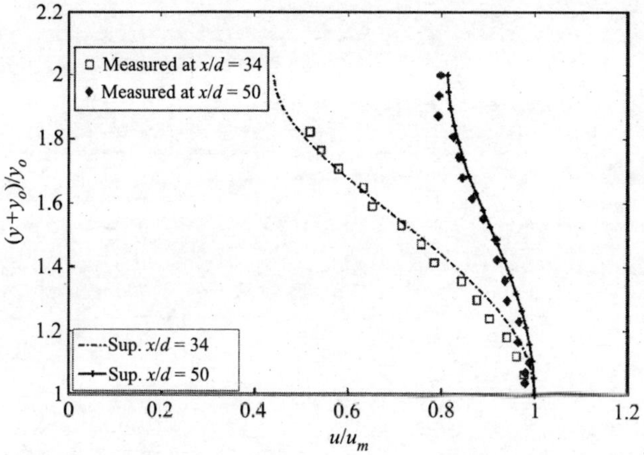

Figure 12.11 Velocity distribution in the surface zone of shallow ambient (Source: Pani and Lai, 2010)

12.7 Summary

The CFD simulation results show that the mean velocity distribution in the fully developed region of a free jet is Gaussian and the profiles satisfy self-similarity. The characteristic length scales b_g and $b_{0.5}$ are linear functions of the distance x and the growth rates are significantly larger than the observed values for all the jets included in this chapter.

The predicted decay of the maximum velocity based on CFD simulation and the point-source technique is nearly the same beyond distances of 10 to 15 times the width of the nozzle. In the near field, differences persist between the two results and the point-source method seems to agree better with the observed data. Overall, the decay of the maximum velocity predicted by the simple point-source method is accurate for all practical purpose.

References

Abramovich, G.N., 1963, *The Theory of Turbulent Jets*, M.I.T. Press, Massachussetts.

Adams, E., 1982, Dilution analysis for unidirectional diffusers, *J. Hyd. Engg.*, 108 (HY3), pp. 3217–342.

Adams, E., Skamaroch, W. and Nothaft, R., 1982, Shoreline effects on the mixing of a Tee diffuser, *J. Hyd. Engg.*, 108 (HY10), pp. 1232.

Albertson, M.L., Dai, Y.B., Jensen, R.A. and Rouse, H., 1950, Diffusion of submerged jets, *Trans. A.S.C.E.*, 115, pp. 639–697.

Ali, K.A.M., 1996; Three dimensional turbulent wall jets in a cross-flow, Ph.D. thesis, Civ. Engg. Dept., I.I.T. Bombay, Mumbai, India.

Anderson, E.A. and Spall, R.E., 2001, Experimental and numerical investigation of two dimensional parallel jets, *Tran. ASME, J. Fluid Engg.* 123, pp. 401–406.

Antonia, R. A. and Bilger, R. W., 1973, An experimental investigation of an axisymmetric jet in a coflowing air stream, *J. Fluid Mech.*, 61, pp. 805–822.

Antonia, R. A. and Bilger, R. W., 1974, The prediction of the axisymmetric turbulent jet issuing into a coflowing stream, *Aeronaut. Q.*, 251, pp. 69–80.

Anwar, H.O. and Weller, J.A., 1969, A thin rectangular jet discharging into slow moving stream, *Water power*, 21, pp. 214–219.

Biringen, S., 1986, An experimental investigation of a turbulent round jet in a coflowing airstream, Proc. ASME Winter Annual Meeting, Anaheim, Calif.

Bradbury, L.J.S., 1965, The structure of a self-preserving turbulent jet, *J. Fluid Mech.*, 23, pp. 31–64.

Chu, V.H., 1994, Lagrangian scalings of jets and plumes with dominant eddies, Recent research advances in the fluid mechanics of turbulent jets and plumes, NATO ASI Series E: Applied Sciences, Vol. 25, P.A. Davies and M.J. Valente Neves, eds., Kluwer, Dordrecht, The Netherlands, pp. 45–72.

Chu, P.C.K., 1996, *Mixing of Turbulent Advected Line Puffs*, Ph.D. thesis, Dept. of Civil and Structural Engineering, University of Hong Kong.

Chu, P.C.K, Lee, J.H.W., and Chu, V.H., 1999, Spreading of turbulent round jet in coflow, *J. Hyd. Engg.*, ASCE, 125(2), pp. 193–204.

Daily, J.W. and Harleman, D.R.F., 1966, Fluid Dynamic, Addiso-Wesley, Reading, Mass.

Dash, R.N., 1981, Studies on single and multiple three dimensional turbulent jets, Civil Engg. Dept., IIT Bombay, Mumbai, India.

Davidson, M.J., 1989, The behaviour of single and multiple, horizontally discharged, buoyant flows in a non-turbulent coflowing ambient fluid, Ph.D. thesis, University of Canterbury, Christchurch.

Davidson, M. J. and Wang, H. J., 2002, Strongly Advected Jet in a Coflow, *J. Hyd. Engg., ASCE*, 128 (8), pp. 742–752.

Demissie, M. and Maxwell, W.H.C., 1982, Three dimensional slot jets, *J. Hyd. Div., ASCE*, 108 (2), pp. 247–251.

Fabis, G. and Fejer, A.A., 1974, Confined mixing of multiple jets, *J. Fluids Engg.*, ASME, Vol., No. 6.

FHWA Urban Drainage Design Manual, 2009, Third Edition, Pub. No. NHI-10-009.

Flora Jr., J.J. and Goldschmidt, V.W., 1969, Virtual origins of a free plane turbulent jet, *J. AIAA*, 7, pp. 2344–2346.

Forstall, W. and Shapiro, A.S., 1950, Momentum and mass transfer in coaxial gas jets, *Trans. ASME, J. Appl. Mech.*, pp. 399–406.

Fujisawa, N. and Shirai, H., 1989, Mean flow and turbulence characteristics of three-dimensional wall jet along plane surface, *Trans. Japan Soc. Aerospace Sci.*, Vol. 32, pp. 35–46.

Glauret, M.B., 1956, The wall jet, *J. Fluid Mech.*, 1, pp. 625–643.

Guymer, I., Dennis, P. and O'Brien, R., 2005, Diameter and surcharge effects on solute transport across surcharged manholes, *J Hyd. Engg., ASCE*, Vol. 131, No. 4, pp. 312–321.

Hakem, M., Hazzab, A. and Ghenaim, A., 2007, Experimental investigation of elliptical jet in coflow, *Int. J. Appl. Engg. Res.*, Jan. 1.

Hinze, J.O. and Zijnen, B.G. Van der Hegge, 1949, Transfer of heat and matter in the turbulent mixing zone of an axially symmetrical jet, *J. Appl. Sci. Res. A1*, pp. 435–461.

Hinze, J.O., 1975, *Turbulence*, McGraw-Hill, New York.

Ho, C.M. and Gutmark, E., 1987, Vortex induction and mass entrainment in a small aspect ratio elliptical jet, *J. Fluid Mech.*, Vol. 179, pp. 383–405.

Hodgson, J.E., Moawad, A.K. and Rajaratnam, N., 1999, Concentration field of multiple circular turbulent jets, *J. Hyd. Res.*, Vol. 37 (2), pp. 249–256.

Hussein, H.J., Capp, S.P. and George, W.K., 1994, Velocity measurement in high-Reynolds-number, momentum-conserving, axisymmetric, turbulent jet, *J. Fluid Mech.*, 258, 31.

Karandikar, J.S., 1997, Hydraulics of duckbill valve jet diffuser, M.Phil. thesis, Civil Engg. Dept., University of Hong Kong, Hong Kong.

Knystautas, R., 1964, The turbulent jet from a series of holes in line, *Aeronautical Quarterly*, Vol. 15, pp. 1–28.

Law, A.W.K. and Herlina, 2002, An experimental study on turbulent circular wall jets, *J. Hyd. Engg., ASCE*, Vol. 28, No. 2, pp. 161–174.

Lee, J.H.W. and Greenberg, M.D., 1984, Line momentum source in shallow inviscid fluid, *J. Fluid Mech.*, Vol. 145, pp. 287–304.

Lee, J.H.W. and Wong, C.F., 1993, Experiments on advected line momentum puffs. Proc. 25th IAHR Congress, Tokyo, Vol. 5, pp. 250–257.

Lee, J.H.W. and Chu, V.H., 2003, *Turbulent jets and plumes – a Lagrangian approach*, Kluwer Academic Press.

Lozanova, M. and Stankov, P., 1998, Experimental investigation on similarity of 3D rectangular turbulent jet, *Exp Fluids*, Vol. 24, issue 5/6, pp. 470–478.

Malmstrom, T.G., Kirkpattrick, A.T., Christensen, K.B. and Knappmiller, K.D., 1997, Centreline velocity decay measurements in low velocity axisymmetric jet, *J. Fluid Mech.*, Vol. 258, 31.

Marsalek, J., 1984, Head losses at sewer junction manholes, *J. Hyd. Engg.*, *ASCE*, Vol. 110, No. 8, pp. 1150–1154.

Maxwell, W.H.C., 1970, Flux development region in submerged jets, *J. Engrg. Mech. Div.*, *ASCE*, 96 (6), pp. 1061–1079.

Mebarka, H., Hazzab, A. and Ghenaim, A., 2007, Experimental investigation of elliptical jet in coflow, *Int. Jr. Applied Engg. Res.*, Vol. 2, No. 1.

Mi, J., Nobes, D.S. and Nathan, G.J., 2001, Influence of jet exit conditions on passive scalar field of an axisymmetric free jet, *J. Fluid Mech.*, 432, pp. 91–125.

Morton, B.R., Taylor, G.I. and Turner, J.S., 1956, Turbulent gravitational convection from maintained and instantaneous sources, *Proc. Royal Soc. London*, Vol. A234, pp. 1–23.

Mudgal, B.V., 1994, Drag characteristics of sills in forced hydraulic jumps and energy loss in junction boxes, Ph.D. thesis, Civil Engineering Department, IIT Bombay, Mumbai, India.

Mudgal, B.V. and Pani, B.S., 1993, Head losses in sewer manholes, Proc. 6th Internation conference on urban storm drainage, Niagra Falls, Ontario, Canada.

Mudgal, B.V. and Pani, B.S., 1992, Drag force on a two-dimensional sill in a wall jet, Proc. 5th Asian Congress of Fluid Mech., Taejon, South Korea.

Myers, G.E., Schauer, J.J. and Eustis, R.H., 1963, The plane turbulent wall jet flow development and friction factor, *J. Basic Engg.*, *Trans. A.S.M.E.*, Vol. 85, pp. 47–54.

Narain, J.P., 1976, Momentum flux development from three dimensional free jets, *J. Fluids Engg.*, *Trans. ASME*, 96 (2), pp. 256–260.

Ozsoy, E., 1977, Flow and mass transport in the vicinity of inlets, Coastal and Oceanographic Engineering Laboratory, Technical Report, No. 36, University of Florida, Gainesville, Florida.

Patil, L.G., 2006, Performance of jet flocculators: A comparative study, Ph.D. thesis, Civil Engineering Department, IIT Bombay, Mumbai, India.

Panchapakesan, N.R. and Lumley, J.L., 1993, Turbulence measurements in axisymmetric jets of air and helium, Part 1, Air Jet, *J. Fluid Mech.*, Vol. 246, 197.

Pande, B.B.L. and Rajaratnam, N., 1975, A predictive method for compound turbulent jets, Technical report, Department of Civil Engineering, University of Alberta, Edmonton.

Pani, B.S., 1972, Three dimensional wall jets, Ph.D. thesis, Civil Engineering Department, University of Alberta, Edmonton, Canada.

Pani, B.S. and Dash, R.N., 1983, Three-dimensional single and multiple free jets, Proc. ASCE, *J. Hyd. Engg.*, 109 (2), pp. 254–269.

Pani, B.S. and Dash, R.N., Three dimensional wall jets from multiple outlets, Proc. Inst. of Civil Engg, London, 75 (2), pp. 735–749.

Pani, B.S. and Parameswaran, P.V., 1994, Momentum and heat flux characteristics of three dimensional jets based on point-source concept, *J. Hyd. Res., IAHR*, 32 (1), pp. 53–66.

Pani, B.S. and Dugad, S.B., 2002, *Turbulent jets: application of point source concept, Research perspectives in hydraulics and water resources engineering*, Ed. Rama Prasad and Vedula, S., World Scientific Pub. Co., Singapore, pp. 1–37.

Pani, B.S., 2006, Jet boundary layer – a point source approach, Proc. HYDRO 2006, organized by the Indian Society for Hydraulics, Pune, India.

Pani, B.S., 2008, Discussion, Interaction of vertical round turbulent buoyant jets- Part II: superposition method, *J. Hyd. Research*, Vol. 46, No. 4, pp. 563–567.

Pani, B.S., Lee, J.H.W. and Lai, A., 2008, *Mixing of coflowing jets in rivers*, 16th IAHR, APD Congress, Nanjing, China, pp. 651–655.

Pani, B.S., 2009, Development of point source method and its practical significance, *Water Sci. Engg.*, Vol. 2, No. 2, pp. 19–31.

Pani, B.S., Lee, J.H.W. and Lai, A.C.H., 2009, Application of Reichardt's hypothesis for multiple coflowing jets, *J. Hyd. Res.*, Vol. 3, pp. 121–128.

Pani, B.S., Lee, J.H.W. and Lai, A.C.H, 2010, *Generalization of Reichardt's Hypothesis: Multiple Coflowing Square Jets*, 17th APD IAHR Congress, Auckland.

Pani, B.S. and Lai, A.C.H., 2010, Investigation of circular jets in shallow water (Discussion on the paper by J.P. Raiford and A.A. Khan)

Pederson, F. and Mark, O., 1990, Head losses in storm sewer manholes; submerged jet theory, *J. Hyd. Engg, ASCE*, Vol. 116, No. 11, pp. 1317–1328.

Parameswaran, P.V., 1986, Charactersitics of three-dimensional heated jets and swirling buoyant jets, Ph.D. thesis, Civil Engineering Department, IIT Bombay, Mumbai, India.

Pratte, B.D. and Baines, W.D., 1967, Profiles of the round turbulent jets in a cross flow, *J. Hyd. Div., Proc. ASCE*, 92, pp. 53–64.

Rajaratnam, N., 1976, *Turbulent Jets*, Elsevier Scientific Publishing Co., Amsterdam.

Rajaratnam, N. and Pani, B.S., 1972, Turbulent compound annular shear layers, Proc. ASCE, *J. Hyd. Div.*, Vol. 98, pp. 1101–1115.

Rajaratnam, N. and Pani, B.S., 1974, Three-dimensional turbulent wall jets, Proc. ASCE, *J. Hyd. Div.*, 100, pp. 69–83.

Rajaratnam, N. and Stalker, M.J., 1982, Circular wall jets in coflowing streams, ASCE, *J. Hyd. Engg.*, Vol. 108, pp. 187–198.

Reichardt, H., 1943, On a new theory of free turbulence, Translated by M. Flint, J. Royal Aeronautical Society, Vol. 47, pp. 167–176 (Originally published in Zamm, 21 (5), 1941).

Sami, S., Carmody, T. and Rouse, H., 1967, Jet diffusion in the region of flow establishment, *J. Fluid Mech.*, Vol. 27, pp. 231–252.

Schlichting, H., 1968, *Boundary Layer Theory*, McGraw-Hill, New York, 6th edition, pp. 747.

Sforza, P.M., Steiger, M.H. and Trentacoste, N., 1966, Studies on three-dimensioanl viscous jets, *J. AIAA*, 4, pp. 800–806.

Sforza, P.M., 1969, A quasi-axisymmetric approximation for turbulent three-dimensional jets and wakes, *J. AIAA*, 7, pp. 1380–1382.

Sigalla, A., 1958, Measurement of skin friction in a plane turbulent wall jet, *J. Royal Aeronautical Soc.*, 62, pp. 873–877.

Sivadas, V., 1994, Experimental investigation on incompressible non-circular turbulent jets in uniform cross flow using laser techniques, Ph.D. thesis, Civil Engineering Department, I.I.T. Bombay, Mumbai, India.

Squire, H.B. and Trouncer, J., 1944, *Round Jets in a General Stream*, Aeronautical Research Council, London, R and M, 1914.

Suresh Kumar N., Pani, B. S. and Joshi, S. G., Sept. 2009, Residual Turbidity in Hydraulic Jet Flocculators, *ISH J. Hydraul. Eng.*, Vol. 15, No. 2, pp. 108–117.

Trentacoste, N. and Sforza, P.M., 1967, Further experimental results for three dimensional free jets, *J. AIAA*, 5, pp 885–891.

Uddin, M. and Pollard, A., 2007, Self-similarity of coflowing jets: The virtual origin, *Phys. Fluid*, 19 (6).

Verhoff, A., 1963, The two-dimensional turbulent wall jet with and without an external stream, Rep. 626, Princeton University.

Wang, H.J., 1996, An experimental study of single and multiple jets in a coflowing environment, M.Phil. thesis, The Hong Kong University of Science and Technology, Hong Kong.

Wang, H.J., 2000, Jet interaction in a still or coflowing environment, Ph.D. thesis, The Hong Kong University of Science and Technology, Hong Kong.

Wright, S.J., 1977, Mean behavior of buoyant jet in a cross flow, *J. Hyd. Div., ASCE*, Vol. 103, HY5, pp. 499–513.

Wright, S.J., 1994, The effect of ambient turbulence on jet mixing, Recent research advances in the fluid mechanics of turbulent jets and plumes, *NATO ASI Series E: App. Sci.*, Vol. 25, P.A. Davies and M.J. Valente Neves, eds., Kluwer, Dordrecht, The Netherlands, pp. 13–27.

Wygnanski, I.J. and Fiedler, H., 1969, Some measurements in self-preserving jet, *J. Fluid Mech.*, Vol. 38, No. 3, pp. 577–612.

Yannopolous, P.C. and Noutsopoulos, G., 2006, Interaction of vertical round turbulent buoyant jets, Part II: Superposition method, *J. Hyd. Res.*, 44 (2), pp. 233–248.

Yevdjevich, V.M., 1966, Diffusion of slot jets with finite orifice length-width ratios, *Hydraul. Pap.*, 2, Colorado State University, Fort Collins.

Yoon, S.E., Kim, J.S., and Song, J.I., 2010, An analysis of head loss coefficient at surcharged manholes wit a 90° bend, 17th IAHR- APD Congress, Auckland.

Index